Explorations and Discoveries in Mathematics

Using The Geometer's Sketchpad® Version 5

A Student Centered Approach to Learning Mathematics

Paul Cinco
Gennadiy Eyshinskiy

Explorations and Discoveries in Mathematics
USING THE GEOMETER'S SKETCHPAD® VERSION 5

Authors:

Paul Cinco
Flushing High School, New York City
Steinhardt School of Education
New York University
Hunter College of the City of New York
Touro College

Gennadiy Eyshinskiy
Flushing High School, New York City

ISBN 978 – 1 – 105 – 38905 - 4

Previously published as Explorations and Discoveries in Mathematics Using the Geometer's Sketchpad, Version 4, Volume 1 © 2005

A note of special thanks is given to Michelle Center for assisting in the editing and proofreading of this manuscript.

Publication of this work has been supported by Key Curriculum Press under a grant to the authors and the Flushing High School Sketchpad Resource Center.
For further information about *Explorations and Discoveries in Mathematics, Using The Geometer's Sketchpad® Version 5,* contact Paul Cinco at commonlog@aol.com

EXPLORATIONS AND DISCOVERIES IN MATHEMATICS
TABLE OF CONTENTS

NOTES TO THE TEACHER

Each of the labs contained in this book is intended to present a specific mathematical concept to the students. The instructional objectives for each of the labs are listed below for convenience so that the teacher can easily select which lab is needed to address a particular instructional need. The labs may be used in a variety of ways.

- A teacher may have access to a computer room where Geometer's Sketchpad® Version 5 is installed on each computer terminal and wishes to bring the class there on a particular day to explore one of the concepts. This would be comparable to the science class having a lab component. The class would be able to write a summary of their observations and have several exercises to show mastery of the concepts.

- A teacher may wish to have a computer with the Geometer's Sketchpad® Version 5 installed and LCD projector on a portable stand that may be rolled into a classroom. In this way, the lab may used as a presentation in the development part of a lesson. Each of the 75 sketches that are produced by the lab instructions may be saved to a diskette or CD for this presentation.

- A teacher may wish to address an individual student need and, having the Geometer's Sketchpad® Version 5 installed on the computers in the school's library, send a student there on a student's free period to complete the lab individually.

- Should the student have the Geometer's Sketchpad® Version 5 installed on the computer at his home, the teacher may wish to assign one of the labs as a homework assignment. The student can then use his work to make a presentation to the class.

- A teacher may want his students to preview a concept the day before the lesson is taught. In this way, the focus of the lesson can be the more rigorous proof of the concept since the concept itself is already known.

The Geometer's Sketchpad® Version 5 Instructional Objectives for each lab are as follows: At the end of the lab, the student will be able to:

Lab #1: The Midsegment of a Triangle

1. State that the midsegment of a triangle that joins the midpoints of any two sides of that triangle is equal to half measure of its third side.
2. Solve arithmetic and algebraic problems involving the midsegment of a triangle.

Lab #2: The Sum of the Angles of a Triangle

1. State that if an angle is designated by three letters, the middle letter is the vertex.
2. State that the sum of the three angles of a triangle equal 180°.
3. Solve arithmetic and algebraic problems involving the three angles of a triangle.

Lab #3: The Angle Bisector

1. Define a ray.
2. Define an angle bisector of an angle.
3. Name two congruent angles in a diagram containing an angle bisector.
4. Solve arithmetic and algebraic problems involving an angle bisector.

Lab #4: The Midpoint of a Line Segment

1. Define a midpoint of a line segment.
2. Name two congruent segments in a diagram containing a midpoint of a line segment.
3. Solve arithmetic and algebraic problems involving a midpoint of a line segment.

Lab #5: The Perpendicular Bisector of a Line Segment

1. Define a perpendicular line.
2. Define a perpendicular bisector of a line segment.
3. Identify a right angle in a diagram containing a perpendicular bisector.
4. Solve arithmetic and algebraic problems involving a perpendicular bisector of a line segment.

Lab #6: The Perimeter of a Polygon

1. Define the perimeter of a triangle, quadrilateral or any polygon.
2. Solve arithmetic and algebraic problems involving perimeter of a polygon.

Lab #7: The Altitude of a Triangle

1. Define an altitude of a triangle.
2. Identify all right angles in a diagram containing an altitude in a triangle.
3. Name perpendicular line segments in a diagram containing an altitude in a triangle.
4. Solve arithmetic and algebraic problems involving an altitude in a triangle.

Lab #8: The Medians of a Triangle

1. Define a median in a triangle.
2. Name two congruent segments in a triangle containing one of its medians.
3. Draw a triangle containing all three of its medians.
4. State that the three medians of a triangle all pass through the same point inside the triangle.
5. Solve arithmetic and algebraic problems involving a median drawn in a triangle.

Lab #9: Complementary Angles

1. Define a pair of complementary angles.
2. Identify the right angle that is formed by two adjacent complementary angles.
3. Identify the perpendicular lines that are formed by two adjacent complementary angles.
4. Solve arithmetic and algebraic problems involving complementary angles.

Lab #10: Supplementary Angles

1. Define a pair of supplementary angles.
2. Identify the straight angle that is formed by two adjacent supplementary angles.
3. Solve arithmetic algebraic problems involving supplementary angles.

Lab #11: Vertical Angles

1. Define a pair of vertical angles.
2. Identify a pair of vertical angles when given two intersecting lines.
3. Solve arithmetic and algebraic problems involving vertical angles.

Lab #12: Parallel Lines – Corresponding Angles

1. Define parallel lines.
2. Identify a transversal that intersects two parallel lines.
3. Identify a pair of corresponding angles in a diagram containing parallel lines.
4. State that when two parallel lines are crossed by a transversal, the corresponding angles are congruent.
5. Solve arithmetic and algebraic problems involving parallel lines and corresponding angles.

Lab #13: Parallel Lines – Alternate Interior Angles

1. Identify a pair of alternate interior angles in a diagram containing parallel lines.
2. State that when two parallel lines are crossed by a transversal, the alternate interior angles are congruent.
3. Solve arithmetic and algebraic problems involving parallel lines and alternate – interior angles.

Lab #14: Parallel Lines – The Interior Angles on the Same Side of the Transversal

1. Identify a pair of interior angles on the same side of the transversal.
2. State that when two parallel lines are crossed by a transversal, the interior angles on the same side of the transversal are supplementary.
3. Solve arithmetic and algebraic problems involving parallel lines and interior angles on the same side of the transversal.

Lab #15: The Exterior Angle of a Triangle

1. Define an exterior angle of a triangle.
2. Identify the remote interior angles relative to the exterior angle.
3. Explain the numerical relationship between the exterior angle of a triangle and its remote interior angles.
4. Solve arithmetic and algebraic problems involving exterior angles of a triangle.

Lab #16: The Sum of the Interior Angles of a Quadrilateral

1. State that the sum of the interior angles of a quadrilateral is 360°.
2. Solve arithmetic and algebraic problems involving the sum of the interior angles of a quadrilateral.

Lab #17: The Sum of the Interior Angles of a Polygon

1. State that the sum of the interior angles of a pentagon is 540°.
2. State the formula $180(n-2)$ for finding the sum of the interior angles of a polygon in terms of the number of sides n of a polygon.
3. Given the number of sides of a polygon, find the sum of its interior angles.
4. Given the sum of the interior angles of the polygon, find the number of sides.
5. Solve arithmetic and algebraic problems involving the sum of the interior angles of a polygon.

Lab #18: The Sum of the Exterior Angles of a Triangle

1. State that the sum of the exterior angles of a triangle is 360°.
2. Solve arithmetic and algebraic problems involving the sum of the exterior angles of a triangle.

Lab #19: The Sum of the Exterior Angles of a Quadrilateral

1. State that the sum of the exterior angles of a quadrilateral is 360°.
2. Solve arithmetic and algebraic problems involving the sum of the exterior angles of a quadrilateral.

Lab #20: The Sum of the Exterior Angles of a Polygon

1. State that the sum of the exterior angles of a polygon is 360° regardless of the number of the sides of a polygon.
2. Solve arithmetic and algebraic problems involving the sum of the exterior angles of a polygon.

Lab #21: Properties of an Isosceles Triangle

1. Define an isosceles triangle in terms of its sides.
2. State that the base angles of an isosceles triangle are congruent.
3. Solve arithmetic and algebraic problems involving the sides or angles of an isosceles triangle.

Lab #22: Properties of an Equilateral Triangle

1. Define an equilateral triangle in terms of its sides.
2. State that each angle of an equilateral triangle contains 60°.
3. Solve arithmetic and algebraic problems involving the sides or angles of an equilateral triangle.

Lab #23: The Triangle Inequality

1. State that the sum of two sides of a triangle must exceed the third.
2. State that if the sum of two sides of a triangle equals the third side, that the triangle degenerates into a line segment.
3. Given the lengths of three sides, identify if it is possible to form a triangle.
4. Given two sides of a triangle, identify all possible values for the third side.

Lab #24: The Pythagorean Theorem

1. Identify that the Pythagorean Theorem applies only to right triangles.
2. Given two sides of a right triangle, to find the remaining side.
3. Solve arithmetic and algebraic problems involving the Pythagorean Theorem.

Lab #25: The Properties of Regular Polygons

1. Define a regular polygon as a polygon having all sides and all angles congruent.
2. Identify an equilateral triangle and square as being members of the set of regular polygons.
3. Solve arithmetic and algebraic problems involving the sides of a regular polygon.

Lab #26: The Opposite Angles of a Parallelogram

1. Define a parallelogram as a quadrilateral with two pairs of opposite sides parallel.
2. State that both pairs of opposite angles of a parallelogram are congruent.
3. Solve arithmetic and algebraic problems involving the opposite angles of a parallelogram.

Lab #27: The Consecutive Angles of a Parallelogram

1. State that any pair of consecutive angles of a parallelogram is supplementary.
2. Solve arithmetic and algebraic problems involving the consecutive angles of a parallelogram.

Lab #28: The Opposite Sides of a Parallelogram

1. State that the opposite sides of a parallelogram are congruent.
2. Solve arithmetic and algebraic problems involving the opposite sides of a parallelogram.

Lab #29: The Diagonals of a Parallelogram

1. State that the diagonals of a parallelogram are not necessarily congruent.
2. State that the diagonals of a parallelogram bisect each other.
3. State that the diagonals of a parallelogram do not necessarily bisect the angles of the parallelogram.
4. Solve arithmetic and algebraic problems involving the diagonals of a parallelogram.

Lab #30: The Properties of a Rectangle

1. Define a rectangle.
2. State that the opposite sides of a rectangle are congruent.
3. State that the four angles of a rectangle are right angles.
4. Solve arithmetic and algebraic problems involving the sides of a rectangle and the angles of a rectangle.

Lab #31: The Diagonals of a Rectangle

1. State that the diagonals of a rectangle are congruent.
2. State that the diagonals of a rectangle bisect each other.
3. State that each diagonal forms a pair of complementary angles at each vertex of the rectangle.
4. Solve arithmetic and algebraic problems involving the diagonals of a rectangle.

Lab #32: The Properties of a Square

1. Define a square as a quadrilateral having four equal sides and at least one right angle.
2. Solve arithmetic and algebraic problems involving the properties of a square.

Lab #33: The Diagonals of a Square

1. State that the diagonals of a square are congruent.
2. State that the diagonals of a square bisect each other.
3. State that the diagonals of a square are perpendicular.
4. State that the diagonals of a square bisect the right angles of a square.
5. Solve arithmetic and algebraic problems involving the properties of a square.

Lab #34: The Properties of a Rhombus

1. Define a rhombus as a quadrilateral having four congruent sides.
2. State that the opposite angles of a rhombus are congruent.
3. Solve arithmetic and algebraic problems involving these properties of a rhombus.

Lab #35: The Diagonals of a Rhombus

1. State that the diagonals of a rhombus are perpendicular.
2. State that the diagonals of a rhombus bisect each other.
3. State that the diagonals of a rhombus bisect the angles of the rhombus.
4. Solve arithmetic and algebraic problems involving the diagonals of a rhombus.

Lab #36: The Properties of a Trapezoid

1. State that a trapezoid is a quadrilateral with only one pair of opposite sides parallel.
2. State that the non-parallel sides are not necessarily congruent.
3. State that the sum of an upper angle and its adjacent lower angle are supplementary.

4. State that the sum of the four angles of a trapezoid is 360°.
5. Solve arithmetic and algebraic problems involving the trapezoid.

Lab #37: The Properties of an Isosceles Trapezoid

1. State that, in an isosceles trapezoid, the non-parallel sides are congruent.
2. State that the upper pair of base angles are congruent and that the lower pair of base angles are congruent.
3. Solve arithmetic and algebraic problems involving the properties of an isosceles trapezoid.

Lab #38: The Diagonals of an Isosceles Trapezoid

1. State that the diagonals of an isosceles trapezoid are congruent.
2. State that the diagonals of an isosceles trapezoid do not bisect each other.
3. Solve arithmetic and algebraic problems involving the diagonals of an isosceles trapezoid.

Lab #39: The Corresponding Angles of Similar Triangles

1. State that the three pairs of corresponding angles in similar triangles are congruent.
2. State that the three pairs of corresponding sides in similar triangles are not necessarily congruent.
3. Solve arithmetic and algebraic problems involving the corresponding angles of similar triangles.

Lab #40: The Corresponding Sides of Similar Triangles

1. State that the ratio of each pair of corresponding sides of a pair of similar triangles is equal.
2. Solve arithmetic and algebraic problems involving the ratios of the corresponding sides of similar triangles.

Lab #41: The Corresponding Perimeters of Similar Triangles

1. State that the ratio of the corresponding perimeters of a pair of similar triangles is the same as the ratio of any of its corresponding sides.
2. Solve arithmetic and algebraic problems involving the ratios of the perimeters of a pair of similar triangles.

Lab #42: The Corresponding Areas of Similar Triangles

1. State the relationship between the ratio of the corresponding sides of a pair of similar triangles and the ratio of the corresponding areas.
2. Given the ratio of corresponding sides of two similar triangles, find the ratio of corresponding areas, and vice versa.
3. Solve arithmetic and algebraic problems involving the ratios of the areas of a pair of similar triangles.

Lab #43: Congruent Triangles: S.S.S. ≅ S.S.S.

1. Define congruent triangles.
2. State that if two triangles agree in S.S.S. ≅ S.S.S., that the corresponding pairs of angles are congruent.
3. Explain why A.A.A. ≅ A.A.A. does not necessarily produce a pair of congruent triangles.
4. Decide if two triangles are congruent based on the S.S.S. criteria.
5. Solve arithmetic and algebraic problems involving corresponding sides of congruent triangles.

Lab #44: Congruent Triangles: S.A.S. ≅ S.A.S.

1. State that if two triangles agree in S.A.S. ≅ S.A.S., all of the other pairs of corresponding parts are congruent.
2. Decide if two triangles are congruent based on the S.A.S. criteria.
3. Solve arithmetic and algebraic problems involving corresponding sides and angles of congruent triangles.

Lab #45: Congruent Triangles: A.S.A ≅ A.S.A.

1. State that if two triangles agree in A.S.A ≅ A.S.A, all of the other pairs of corresponding parts are congruent.
2. Decide if two triangles are congruent based on the A.S.A. criteria.
3. Solve arithmetic and algebraic problems involving corresponding sides and angles of congruent triangles.

Lab #46: A Reflection in a Line

1. Describe a line reflection in a plane.
2. Given a polygon, draw its reflection in a given line.
3. State that the image of a polygon after a reflection in a line produces a polygon congruent to the original polygon.
4. State that the distance from any point on a polygon to its line of reflection is the same on either side of this line.
5. Solve arithmetic and algebraic problems involving line reflections of polygons and their images.

Lab #47: A Translation in the Plane

1. Describe a translation in the plane.
2. Given a polygon, draw its image after a given translation.
3. State that the image of a polygon after a translation produces a polygon congruent to the original polygon.
4. State that the distance from any point on a polygon to the image of the point is the same.
5. Solve arithmetic and algebraic problems involving translation in the plane of polygons and their images.

Lab #48: A Rotation about a Point

1. Describe a rotation about a point.
2. Given a polygon and an angle of rotation, draw its image after a rotation about a given point.
3. State that the image of a polygon after a rotation produces a polygon congruent to the original polygon.
4. State that the angle formed between a specific point on a polygon and its image about the point of rotation is the angle of rotation.
5. Solve arithmetic and algebraic problems involving rotations.

Lab #49: A Dilation about a Point in the Plane

1. Describe a dilation about a point in the plane.
2. Given a point and a constant of dilation, draw the image of a polygon under a dilation about this point.
3. State that the image of a polygon after a dilation produces a polygon similar to the original polygon.
4. Given a pair of corresponding sides of a polygon and its image after a given dilation, find the constant of dilation.
5. Solve arithmetic and algebraic problems involving dilations.

Lab #50: The Median of a Trapezoid

1. Define the median of a trapezoid.
2. Given a trapezoid, draw its median.
3. Given the upper and lower base of a trapezoid, compute the length of its median.
4. Given the length of a median and one of the trapezoid's bases, find the other base.
5. Solve arithmetic and algebraic problems involving the median of a trapezoid.

Lab #51: The Area of a Square

1. State the formula for the area of a square in terms of a side.
2. Given a side of a square, find the area.
3. Given the area of a square, find a side.
4. Solve arithmetic and algebraic problems involving the area of a square.

Lab #52: The Area of a Rectangle

1. State the formula for the area of a rectangle in terms of its sides.
2. Given the length and width of a rectangle, find its area.
3. Given the area and one side, to find all remaining sides of a rectangle.
4. Solve arithmetic and algebraic problems involving the area of a rectangle.

Lab #53: The Area of a Parallelogram

1. State the formula for the area of a parallelogram in terms of its base and altitude.
2. State that the altitude is not the same as one of the sides of the parallelogram.
3. Given the base and altitude of a parallelogram, find its area.
4. Given the area and either its base or altitude of a parallelogram, find the missing base or altitude.
5. Solve arithmetic and algebraic problems involving the area of a parallelogram.

Lab #54: The Area of a Triangle

1. State the formula for the area of a triangle in terms of its base and altitude.
2. State that the altitude is not necessarily the same as one of the sides of the triangle.
3. Given the base and altitude of a triangle find its area.
4. Given the area and either the base or altitude of a triangle, find the missing base or altitude.
5. Solve arithmetic and algebraic problems involving the area of a triangle.

Lab #55: The Area of a Trapezoid

1. State the formula for the area of a trapezoid in terms of its upper and lower bases and altitude.
2. Given the bases and altitude of a trapezoid, find the area.
3. Solve arithmetic and algebraic problems involving the area of a trapezoid.

Lab #56: The Area of a Square Using its Diagonals

1. State the formula for the area of a square in terms of its diagonals.
2. Given the diagonal of a square, find the area.
3. Given the area of a square, find the diagonal.
4. Solve arithmetic and algebraic problems involving the area of a square in terms of its diagonals.

Lab #57: Distinguishing Between the Area and Perimeter of a Triangle

1. Explain the difference between the area and perimeter of a triangle.
2. Solve arithmetic and algebraic problems involving distinguishing the difference between the area and perimeter of a triangle.

Lab #58: The Radius and Diameter of a Circle

1. Define the radius and diameter of a circle.
2. Given the radius of a circle, find the diameter.
3. Given the diameter of a circle, find the radius.
4. Solve arithmetic and algebraic problems involving the radius and diameter of a circle.

Lab #59: The Circumference of a Circle and π

1. Define the circumference of a circle.
2. State that the quotient a the circumference of a circle and its diameter is π.
3. State that the formulas $C = \pi d$ or $C = 2\pi r$ can be used to find the circumference of a circle.
4. Find the circumference of a circle given its radius or diameter.
5. Find the diameter or radius of a circle given its circumference.
6. Solve arithmetic and algebraic problems involving the circumference of a circle.

Lab #60: The Area of a Circle

1. State the formula for finding the area of a circle.
2. Find the radius or diameter of a circle given its area.
3. Solve arithmetic and algebraic problems involving the area of a circle.

Lab #61: The Three Angle Bisectors of a Circle

1. State that the three angle bisectors of a triangle are concurrent at a point inside the given triangle.
2. Define the incenter of a triangle.
3. State that the distance between the incenter of a triangle and any of its sides is the same.

Lab #62: The Tangent Ratio in a Right Triangle

1. State the tangent ratio for a given acute angle in a right triangle.
2. State that the tangent ratio for a given acute angle does not depend on the size of the right triangle.
3. Find the tangent of a given acute angle correct to the nearest thousandth if given a right triangle and its two legs.
4. Find the leg opposite a given an acute angle of a right triangle and the leg adjacent the angle.

Lab #63: The Sine Ratio in a Right Triangle

1. State the sine ratio for a given acute angle in a right triangle.
2. State that the sine ratio for a given acute angle does not depend on the size of the right triangle.
3. Find the sine of a given acute angle correct to the nearest thousandth if given a right triangle, the leg opposite the angle and the hypotenuse.
4. Find the hypotenuse of a right triangle if given an acute angle and the leg opposite the angle.

Lab #64: The Cosine Ratio in a Right Triangle

1. State the cosine ratio for a given acute angle in a right triangle.
2. State that the cosine ratio for a given acute angle does not depend on the size of the right triangle.
3. Find the cosine of a given acute angle correct to the nearest thousandth if given a right triangle, the leg adjacent the angle and the hypotenuse.
4. Find the hypotenuse of a right triangle if given an acute angle and the leg adjacent the angle.

Lab #65: Naming a Point by Using its Coordinates

1. Plot a point on a set of coordinate axes if given its coordinates.
2. Read the coordinates of a plotted point.
3. Define abscissa and ordinate for a point.
4. Identify the location of Quadrants I, II, III and IV on a set of coordinate axes.
5. Identify the signs of the abscissa and ordinate of a point in a given quadrants.

Lab #66: The Slope of a LIne

1. State the formula for finding the slope of a line joining two given points.
2. Use the slope formula to find the slope of a line passing through two points whose coordinates are given.

Lab #67: The Slopes of Parallel Lines

1. State that if two lines are parallel, then their slopes are equal.
2. State that if the slopes of two lines are equal, then the lines are parallel.
3. Solve arithmetic and algebraic problems involving the slopes of parallel lines.

Lab #68: The Slopes of Perpendicular Lines

1. State that if two lines are perpendicular, then the product of their slopes is -1.
2. State that the product of the slopes of two lines is -1, that the two lines are perpendicular.
3. Solve arithmetic and algebraic problems involving the slopes of perpendicular lines.

Lab #69: The Slope and the y-intercept of a Line

1. State that a line graphed on a coordinate plane has an equation.
2. State that the equation of a line graphed on a coordinate plane is $y = mx + b$.
3. State that the y-intercept of a line is the number that follows the x-term in the equation written in the standard form.
4. State that slope of a line is the coefficient of the x-term in the equation written in the standard form.
5. Find the slope and y-intercept of a graphed line.
6. Write the equation of a line given its slope and y-intercept.

Lab #70: The Equation of a Vertical Line on the Coordinate Plane

1. State that the abscissa of any vertical line graphed on the coordinate plane is constant.
2. State that the equation of a vertical line graphed on the coordinate plane is in the form $x = c$, where c is a constant.
3. State the equation of the y-axis.
4. Graph a vertical line given an equation in the form of $x = c$.

Lab #71: The Equation of a Horizontal Line on the Coordinate Plane

1. State that the ordinate of any horizontal line graphed on the coordinate plane is constant.
2. State that the equation of a horizontal line graphed on the coordinate plane is in the form $y = c$, where c is a constant.
3. State the equation of the x-axis.
4. Graph a horizontal line given an equation in the form of $y = c$.

Lab #72: The Coordinates of the Midpoint of a Line Segment

1. State and apply the formula for finding the midpoint of a line segment whose coordinates of its endpoints are given.
2. Find the coordinates of one of the endpoints of a line segment when given the coordinates of the midpoint and the coordinates of the other endpoint.

Lab #73: The Distance between the Endpoints of a Horizontal Line Segment

1. State and apply the formula for finding the length of a horizontal line segment given the coordinates of the endpoints.
2. Explain why the order of subtraction in the distance formula does not affect the length of the line segment.
3. Explain why absolute value symbols are part of this distance formula.
4. Find the coordinates of an endpoint of a horizontal line segment when given the length of the segment and the other endpoint.

Lab #74: The Distance Between the Endpoints of a Vertical Line Segment

1. Find the length of a vertical line segment graphed on a set of coordinate axes when given the coordinates of the endpoints.
2. State and apply the formula for finding the length of a horizontal line segment.
3. Explain why the order of subtraction in the distance formula does not affect the length of the line segment.
4. Explain why absolute value symbols are part of this distance formula.
5. Find the coordinates of an endpoint of a vertical line segment when given the length of the segment and the other endpoint.

Lab #75: The Distance Between the Endpoints of Any Line Segments

1. State and apply the distance formula to any line segment graphed on a set of coordinate axes.
2. Explain that the distance formula is an application of the Pythagorean Theorem.
3. Use the distance formula to prove that a triangle is isosceles.

A VERY IMPORTANT NOTE ON SELECTING AND DESELECTING OBJECTS

Objects in a sketch may only be selected and deselected using the Arrow tool.

1. To select an object, such as a point, a line, or a circle, is to left-click on the object. This action will highlight the selected object in the sketch.

2. To deselect a highlighted object in the sketch, left-click anywhere in the blank region of the sketch. This action will cause the object to be non-highlighted.

3. You may not be able to use the [CONSTRUCT] OR [MEASURE] menus if extra objects are selected.

THE

LABS

LAB #1: THE MIDSEGMENT OF A TRIANGLE

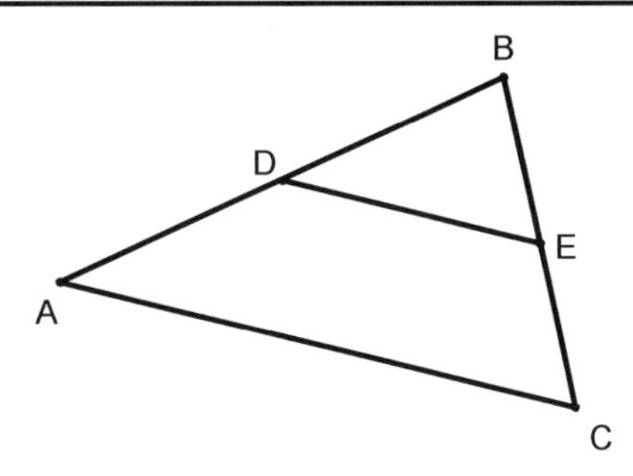

1. Select: [EDIT], [PREFERENCES], [TEXT], check the box [FOR ALL NEW POINTS] and click [OK].
2. Select: the Point tool and plot 3 points randomly in a clockwise direction.
3. Select: the Arrow tool, [EDIT], [SELECT ALL], [CONSTRUCT], [SEGMENTS] and click anywhere in the blank region of your sketch to deselect all segments. (This constructs ΔABC.)
4. Place the cursor/pointer on side $\overline{AB}$ and left click to select it, then place the cursor/pointer on side $\overline{BC}$ and left click to select it. (*Both sides* $\overline{AB}$ *and* $\overline{BC}$ *should now be highlighted!*)
5. Select: [CONSTRUCT], [MIDPOINTS], [CONSTRUCT], [SEGMENT], [MEASURE], [LENGTH] and click anywhere in the blank region of your sketch to deselect the highlighted measure of line segment $\overline{DE}$. (Midsegment $\overline{DE}$ joins midpoints of sides $\overline{AB}$ and $\overline{BC}$ of ΔABC.)
6. Select: side $\overline{CA}$, [MEASURE], [LENGTH], [NUMBER], [CALCULATE], click on the caption that shows $m\overline{CA}$, [DIVISION SYMBOL ON THE CALCULATOR , ÷], [2], [OK] and deselect.
7. Select: the Text tool, double click in the blank region of your sketch to open a dialog box and type your observation of $m\overline{DE}$ and the value of $\frac{m\overline{CA}}{2}$.
8. Select: the Arrow tool and click anywhere in the blank region of your sketch to deselect the highlighted text.

9. Click on vertex A of the triangle and keeping the left button of the mouse depressed, drag it to change the size of the triangle. Observe values that remain equal as you do so.
10. Select: the Text tool, double click in the blank region of your sketch to open a dialog box and (1) write an explanation of how to find the length of the midsegment that joins midpoints of two sides of a triangle if you know the measure of the third side of that triangle, (2) use your explanation to replace the question mark in the expression $m\overline{DE}\ ?\ \frac{m\overline{CA}}{2}$ with the appropriate sign, >, < , or =.

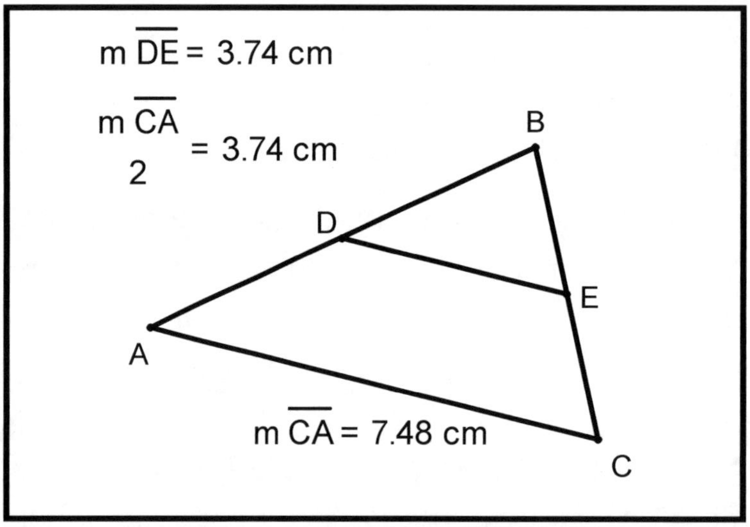

$m\overline{DE} = 3.74$ cm

$\frac{m\overline{CA}}{2} = 3.74$ cm

B

D

E

A

$m\overline{CA} = 7.48$ cm

C

SUGGESTED EXERCISES

Type the solutions to the following problems in the blank region of your sketch.

1. In of ΔABC $m\overline{CA} = 12$ in. Find $m\overline{DE}$.
2. In of ΔABC, $m\overline{CA} = 16$ cm. If $m\overline{DE} = 2x$, find the value of x
3. In of ΔABC $m\overline{CA} = 2(2x + 10)$ and $m\overline{DE} = x + 20$. Find the value of x and the measures of side $\overline{CA}$ and midsegment $\overline{DE}$.
4. In the sketch that you have just made by following the instructions of this lab, construct the midpoint F on side $\overline{CA}$, construct midsegment $\overline{EF}$, measure side $\overline{AB}$, measure midsegment $\overline{EF}$ and verify that that the measure of midsegment $\overline{EF}$ is a half measure of side $\overline{AB}$.

LAB #2: THE SUM OF THE ANGLES OF A TRIANGLE

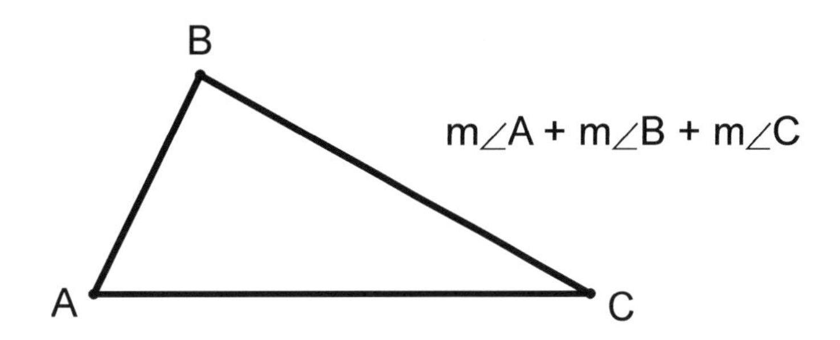

1. Select: [EDIT], [PREFERENCES], [TEXT], check the box [FOR ALL NEW POINTS] and click [OK].
2. Select: the Point tool and plot 3 points randomly as the vertices of a triangle.
3. Select: the Arrow tool, [EDIT], [SELECT ALL], [CONSTRUCT], [SEGMENTS] and click anywhere in the blank region of your sketch to deselect all segments.(YOU HAVE JUST CONSTRUCTED ΔABC.)
4. Select: points A, then B, then C, [MEASURE] and [ANGLE].
5. Click anywhere in the blank region of the sketch to deselect.
6. Select: points B, then C, then A, [MEASURE] and [ANGLE].
7. Click anywhere in the blank region of the sketch to deselect.
8. Select: points B, then A, then C, [MEASURE] and [ANGLE].
9. Click anywhere in the blank region of the sketch to deselect.
10. Select: [NUMBER] and [CALCULATE]. A calculator will appear.
11. Click on the caption that shows m∠ABC, [+], click on the caption that shows m∠BCA, [+], click on the caption that shows m∠BAC and [OK].
12. Click on any vertex of the triangle and keeping the left button of the mouse depressed, drag it to change the size of the triangle. Observe what value remains constant as you do so.

13. Select: the Text tool, double click in the blank region of your sketch to open a dialog box and explain what you have learned about the sum of the angles of the triangle. Use complete sentences to explain what value remained constant as you changed the size of the triangle.

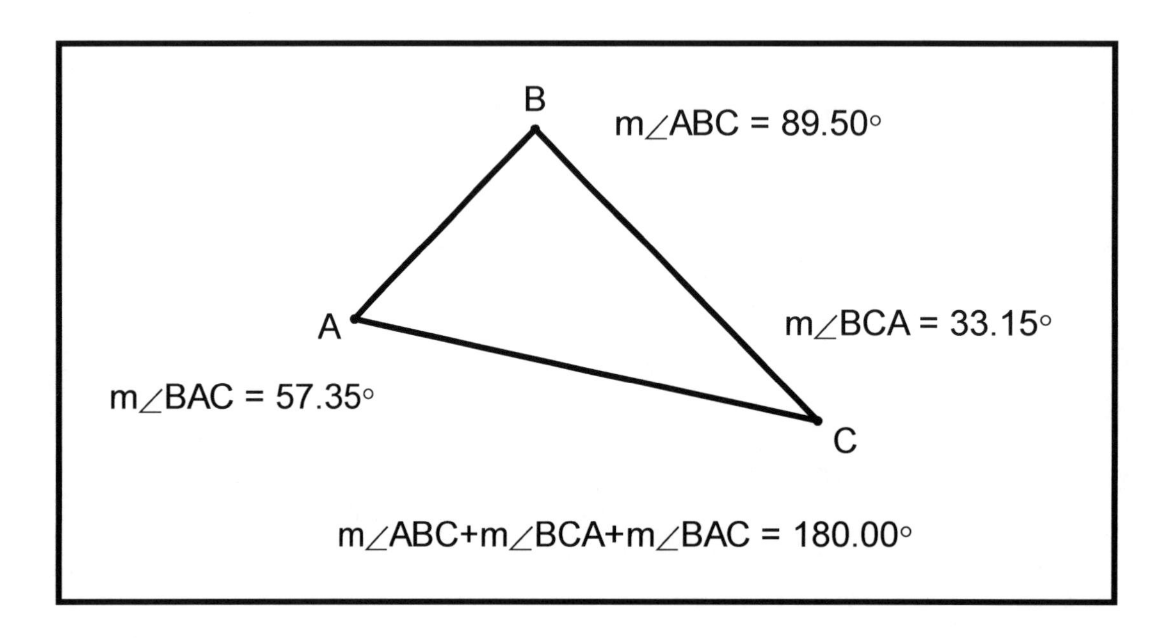

SUGGESTED EXERCISES

Type the solutions to the following problems in the blank region of your sketch.

1. In ΔABC $m\angle A = 43°$ and $m\angle B = 112°$, find $m\angle C$.

2. In ΔABC $m\angle A = 2x$, $m\angle B = 98°$ and $m\angle C = 52°$, find the value of x.

3. In ΔABC $m\angle A = 2x + 10$, $m\angle B = 3x + 20$ and $m\angle C = 45°$, find the value of x, $m\angle A$ and $m\angle B$.

LAB #3: THE BISECTOR OF AN ANGLE

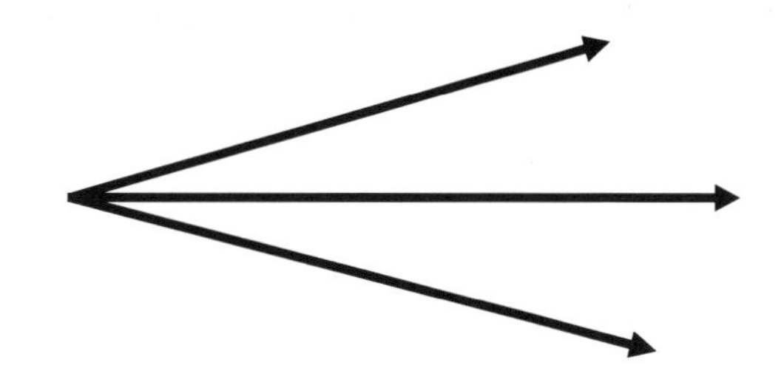

1. Select: [EDIT], [PREFERENCES], [TEXT], check the box [FOR ALL NEW POINTS] and click [OK].
2. Select: the Point tool and plot 3 points that do not lie in the same line.
3. Select: the Arrow tool and click in the blank region of the sketch to deselect.
4. Select: points A, then B, [CONSTRUCT], [RAY] and deselect.
5. Select: points A, then C, [CONSTRUCT], [RAY] and deselect.
6. Select: point B, then A, then C, [MEASURE], [ANGLE], [EDIT], [SELECT PARENTS], [CONSTRUCT], [ANGLE BISECTOR], [CONSTRUCT] and [POINT ON BISECTOR].
7. Click on point D, and keeping the left button depressed, drag it about two inches away from point A and deselect.
8. Select: points B, then A, then D, [MEASURE], [ANGLE] and deselect.
9. Select: points C, then A, then D, [MEASURE], [ANGLE] and deselect.
10. Select: [NUMBER], [CALCULATE], click on the caption that shows $m\angle BAD$, [+], click on the caption that shows $m\angle CAD$ and [OK].
11. Select: the Text tool, double click in the blank region to open a dialog box, and (1) describe your observations about the measures of $\angle BAD$ and $\angle CAD$,
    (2) explain to the measure of what angle their sum is equal.
12. Select: the Arrow tool, click on points A, B or C, and keeping the left button depressed, drag it. Observe the measures that remain equal as you do so.

13. Select: the Text tool, double click in the blank region to open a dialog box and explain what the angle bisector $\overrightarrow{AD}$ is, and what it does to the bisected ∠ BAC.

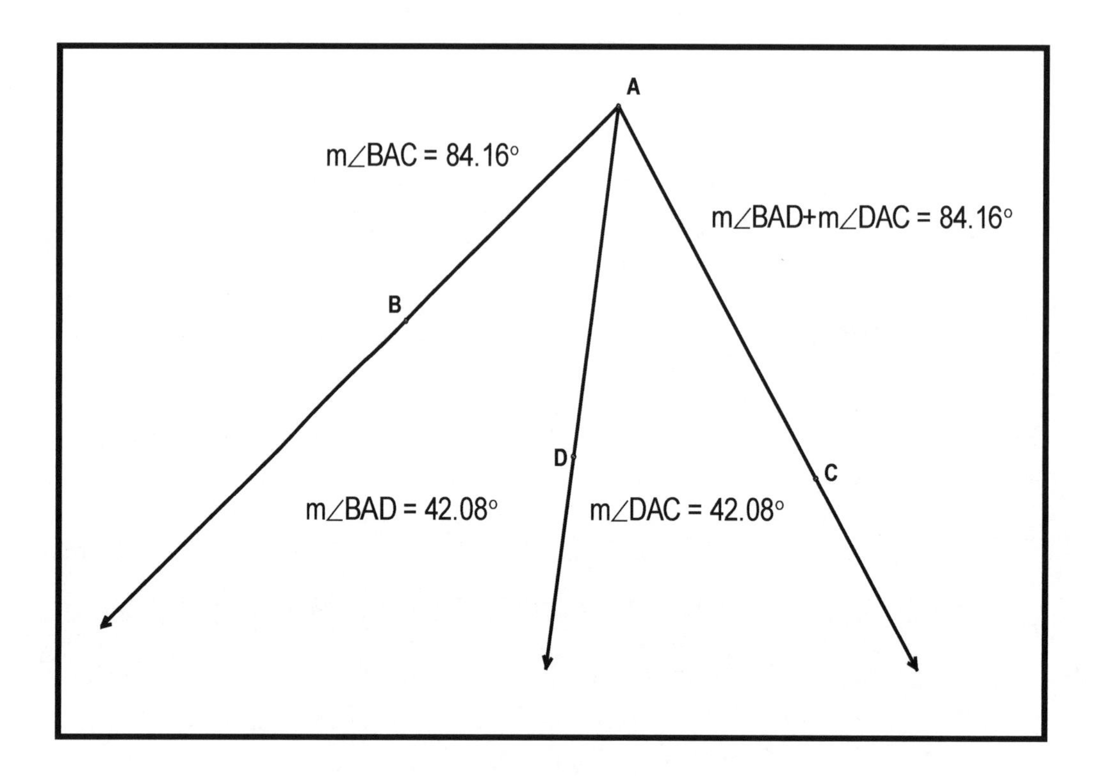

SUGGESTED EXERCISES

Type the solutions to the following problems in the blank region of your sketch.

1. $\overrightarrow{AD}$ is the angle bisector of BAC. Find the measure of ∠BAD if the measure of CAD is 43°.

2. $\overrightarrow{AD}$ is the angle bisector of ∠BAC. Find the value of x if m∠BAD = 3x – 4 and m∠CAD = 38°.

3. $\overrightarrow{AD}$ is the angle bisector of ∠BAC. Find m∠ BAD if m∠CAD = 2x +10 and m∠BAD = 4x.

4. $\overrightarrow{AD}$ is the angle bisector of BAC. Find m∠BAD if m∠BAC = 80°.

LAB #4: THE MIDPOINT OF A LINE SEGMENT

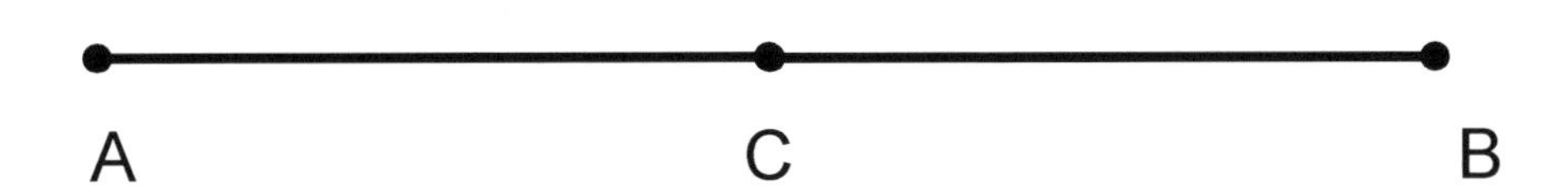

1. Select: [EDIT], [PREFERENCES], [TEXT], check the box [FOR ALL NEW POINTS] and click [OK].
2. Select: the Point tool and plot two points as the endpoints of a horizontal line segment.
3. Select: the Arrow tool, [EDIT], [SELECT ALL], [CONSTRUCT], [SEGMENT], [CONSTRUCT], [MIDPOINT] and deselect.
4. Select: points A, then C, then B, [CONSTRUCT], [SEGMENTS], [MEASURE], [LENGTHS], and deselect.
5. Select: [NUMBER], [CALCULATE], the caption that shows $m\overline{AC}$, [+], the caption that shows $m\overline{CB}$ and [OK].
6. Select: the Text tool, double click to open a dialog box and (1) describe your observations about the measures of line segments $\overline{AC}$ and $\overline{CB}$,

   (2) explain to the measure of what line segment their sum is equal.
7. Select: the Arrow tool, click on point A or B, and keeping the left button depressed, drag it. Observe what values remain equal as you do so.
8. Select: the Text tool, double click in the blank region to open a dialog box and explain what the midpoint C does to $\overline{AB}$. Give your own definition of a midpoint.

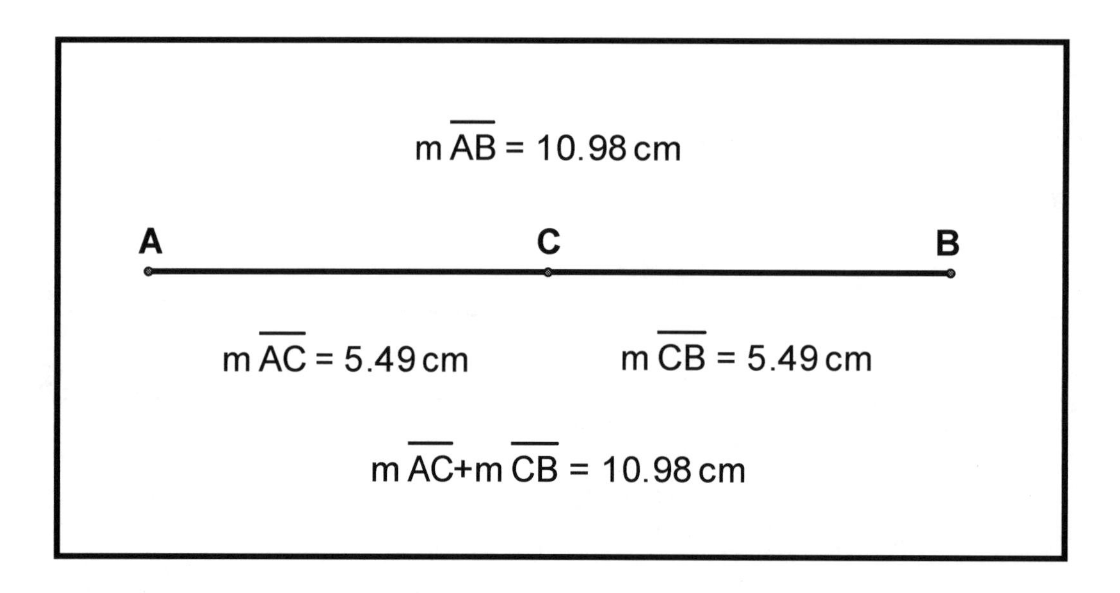

SUGGESTED EXERCISES

Type the solutions to the following problems in the blank region of your sketch.

1. C is the midpoint of $\overline{AB}$. Find the measure of line segment $\overline{AC}$ if the measure of line segment $\overline{BC}$ is 7 cm.

2. C is the midpoint of $\overline{AB}$. Find the value of x if $m\overline{AC} = 2x + 5$ and $m\overline{BC} = 21$ in.

3. C is the midpoint of $\overline{AB}$. If $m\overline{AC} = 3x + 40$ and $m\overline{BC} = x + 80$, find the values of x, the length of line segment $\overline{BC}$, and the length of line segment $\overline{AB}$.

4. C is the midpoint of $\overline{AB}$. If $m\overline{AC} = 2x + 3$, express the measure of line segment $\overline{AB}$ in terms of x.

LAB #5: THE PERPENDICULAR BISECTOR OF A LINE SEGMENT

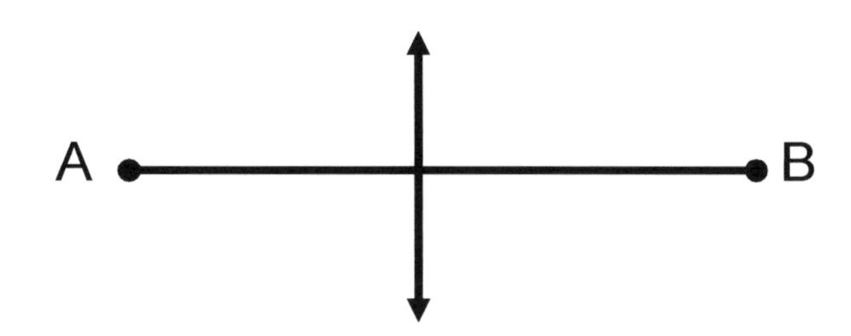

1. Select: [EDIT], [PREFERENCES], [TEXT], check the box [FOR ALL NEW POINTS] and [OK].
2. Select: the Point tool and plot two points as the endpoints of a horizontal line segment.
3. Select: the Arrow tool, [EDIT], [SELECT ALL], [CONSTRUCT], [SEGMENT], [CONSTRUCT], [MIDPOINT] and deselect.
4. Select: line segment $\overline{AB}$, midpoint C, [CONSTRUCT], [PERPENDICULAR LINE], [CONSTRUCT], [POINT ON PERPENDICULAR LINE] and deselect.
5. Select: points A, then C, then B, [CONSTRUCT], [SEGMENTS], [MEASURE], [LENGTHS] and deselect.
6. Click on point A or B, and keeping the left button depressed, drag it. Observe the measures that remain equal as you do so.
7. Select: the Text tool, double click in the blank region to open a dialog box and (1) describe your observations about the measures of $\overline{AC}$ and $\overline{CB}$, (2) explain what a perpendicular bisector drawn through C did to line segment $\overline{AB}$.
8. Select: the Arrow tool and click in the blank region of the sketch to deselect.
9. Select: points A, then C, then D, [MEASURE], [ANGLE] and deselect.
10. Select: points B, then C, then D, [MEASURE], [ANGLE] and deselect.

11. Click on point A or B, and keeping the left button depressed, drag it. Observe the changes, if any, in the values of the measured angles.
12. Select: the Text tool, double click in the blank region to open a dialog box and describe the type of angles the perpendicular bisector drawn through C, forms with line segment $\overline{AB}$.

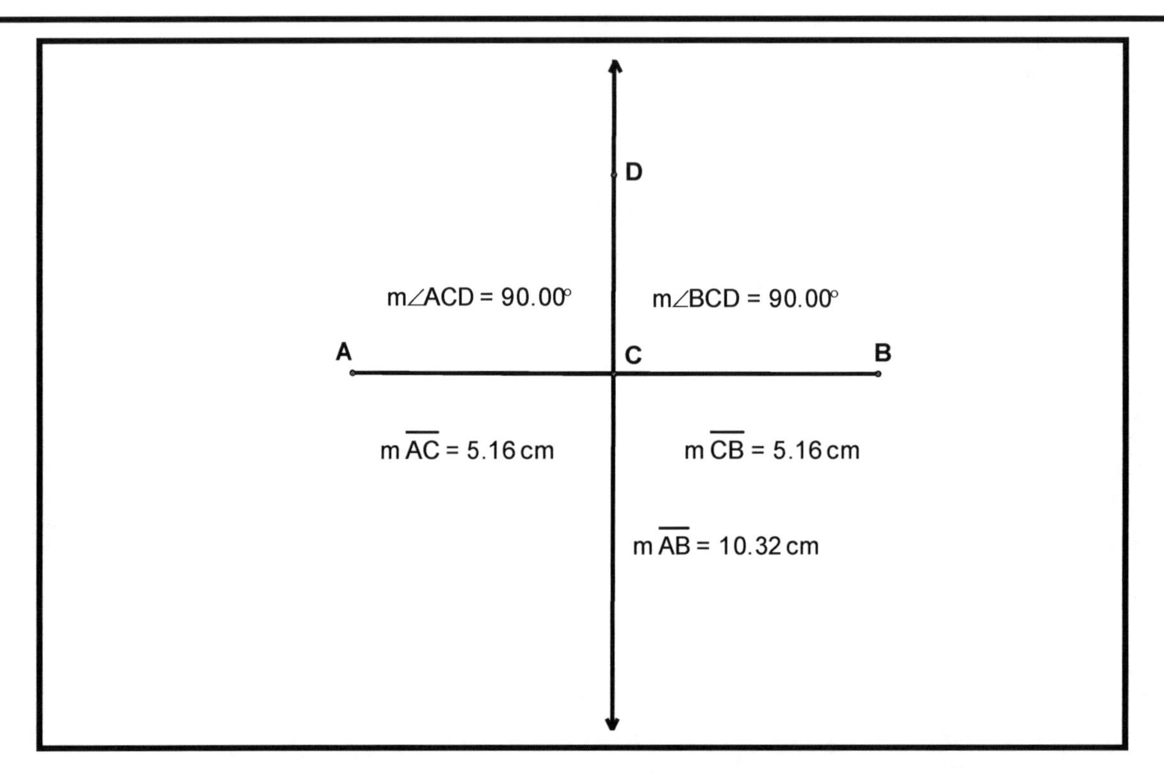

SUGGESTED EXERCISES

Type the solutions to the following problems in the blank region of your sketch.

1. $\overleftrightarrow{CD}$ is the perpendicular bisector of $\overline{AB}$. Find m∠ACD if m∠BCD = 90°.

2. $\overleftrightarrow{CD}$ is the perpendicular bisector of $\overline{AB}$. Find the value of x if m∠ACD = 5x.

3. $\overleftrightarrow{CD}$ is the perpendicular bisector $\overline{AB}$. Find the measure of $\overline{AC}$ if m$\overline{BC}$ = 3.22 cm.

4. $\overleftrightarrow{CD}$ is the perpendicular bisector $\overline{AB}$. Find the value of x if m$\overline{AC}$ = 4x + 2 and m$\overline{BC}$ = 14 cm.

LAB #6: THE PERIMETER OF A POLYGON

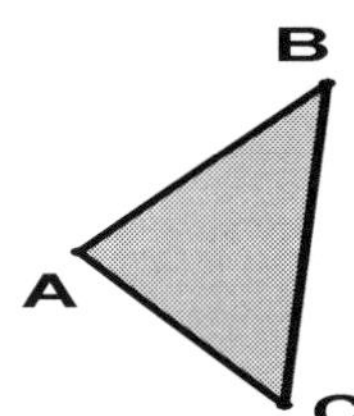

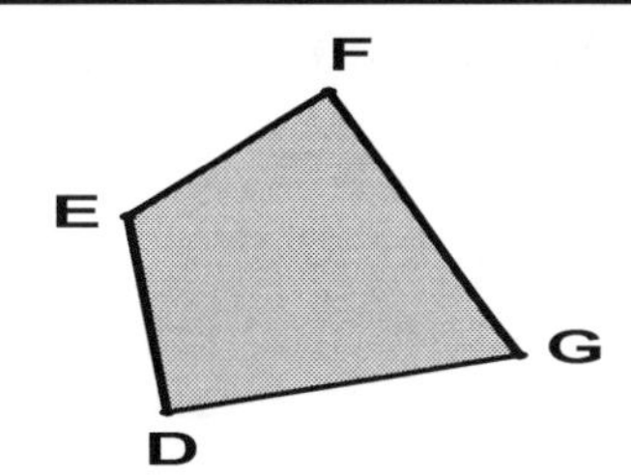

1. Select: the Point tool and plot three points on the left part of your screen as the vertices of a triangle.
2. Select: the Arrow tool, [EDIT], [SELECT ALL], [CONSTRUCT], [SEGMENTS], [MEASURE], [LENGTHS] and deselect.
3. Select: [NUMBER], [CALCULATE], click on the caption that shows $m\overline{AB}$, [+], click on the caption that shows $m\overline{BC}$, [+], click on the caption that shows $m\overline{CA}$, [OK] and deselect.
4. Select: points A, then B, then C, [CONSTRUCT], [TRIANGLE INTERIOR], [MEASURE] and [PERIMETER].
5. Select: the Text tool, double click to open a dialog box and describe your observations about the sum of the sides $m\overline{AB} + m\overline{BC} + m\overline{CA}$ and the measure of the perimeter of ΔABC.
6. Select: the Point tool and plot four points in the <u>clockwise direction</u> on the right part of your screen as the vertices of a quadrilateral.
7. Select: the Arrow tool and click in the blank region to deselect.
8. Select: all four newly plotted points in the <u>clockwise direction</u>, [CONSTRUCT], [SEGMENTS], [MEASURE], [LENGTHS] and deselect.
9. Select: [NUMBER], [CALCULATE], click on the caption that shows $m\overline{DE}$, [+], click on the caption that shows $m\overline{EF}$, [+], click on the caption that shows $m\overline{FG}$, [+], the caption that shows $m\overline{GD}$, [OK] and deselect.
10. Select: point D, then E, then F, then G, [CONSTRUCT], [QUADRILATERAL INTERIOR], [MEASURE] and [PERIMETER].
11. Select: the Text tool, double click in the blank region to open a dialog box and describe your observations about the sum of the sides

$m\overline{DE} + m\overline{EF} + m\overline{FG} + m\overline{GD}$ and the measure of the perimeter of the quadrilateral DEFG.

12. Select: the Arrow tool, click on any labeled point of ΔABC and, keeping the left button of the mouse depressed, drag it. Observe the values that remain equal as you do so.
13. Do the same to drag any labeled point of the quadrilateral DEFG Observe the values that remain equal as you do so.
14. Select: the Text tool, double click in the blank region to open the dialog box and explain what you have to do with the sides of the geometric figure to find its perimeter. Give your own definition of the perimeter.

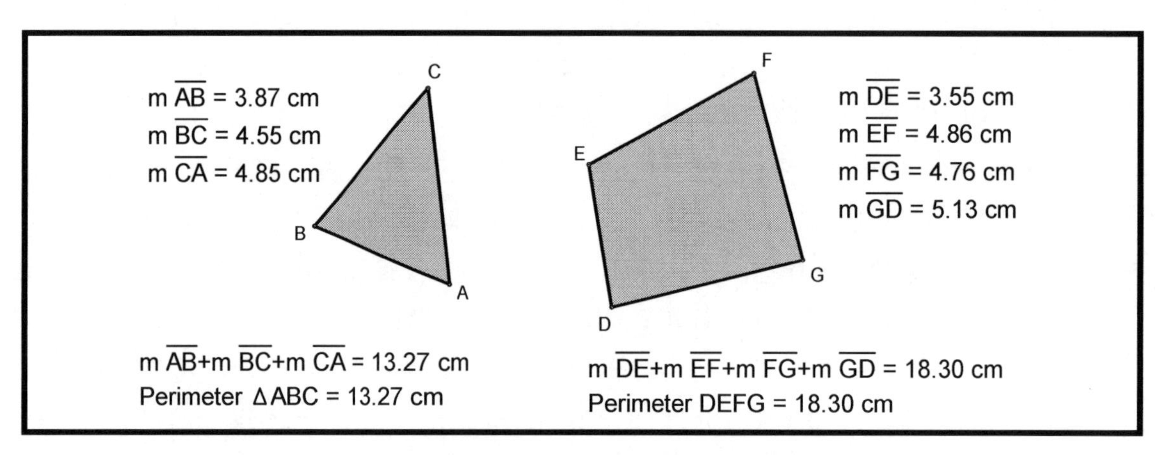

SUGGESTED EXERCISES

Type the solutions to the following problems in the blank region of your sketch.

1. Find the perimeter of a triangle whose sides are 3 in, 5 in, and 6 in.

2. The perimeter of ΔABC is 17 in. Find the measure of side $\overline{AB}$ if $m\overline{BC} = 8$ in and $m\overline{AC} = 5$ in.

3. The perimeter of quadrilateral DEFG is 30 in. If $m\overline{DE} = 4$ in, $m\overline{EF} = 7$ in, $m\overline{FG} = 3$ in, and $m\overline{DG} = 2x$, find the value of x and the measure of the side $\overline{DG}$.

4. The perimeter of ΔABC is 20 cm. If $m\overline{AB} = 2x + 4$, $m\overline{BC} = 3x - 2$, and $m\overline{AC} = x + 6$, find the value of x and the measure of each side.

LAB #7: THE ALTITUDE OF A TRIANGLE

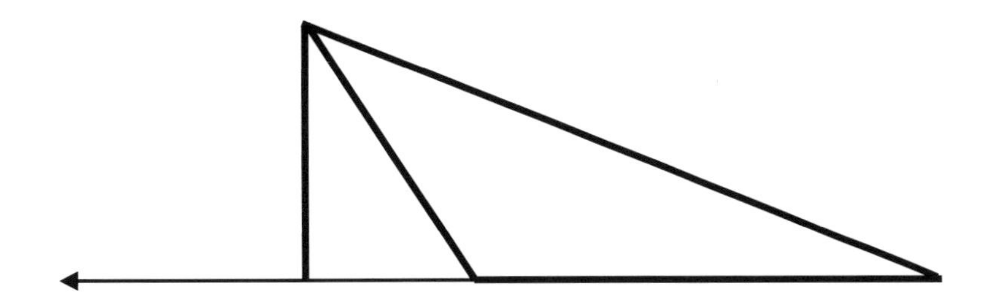

1. Select: [EDIT], [PREFERENCES], [TEXT], check the box [FOR ALL NEW POINTS] and click [OK].
2. Select: the Point tool and plot two points as the endpoints of a horizontal line segment.
3. Select: the Arrow tool, [EDIT], [SELECT ALL], [CONSTRUCT], [LINE], [DISPLAY], [LINE STYLE], [THIN], [DISPLAY], [LINE STYLE] and [DASHED].
4. Select: the Point tool and plot a point anywhere above line $\overleftrightarrow{AB}$.
5. Select: the Arrow tool and click in the blank region to deselect.
6. Select: point C, line $\overleftrightarrow{AB}$, [CONSTRUCT], [PERPENDICULAR LINE] and deselect.
7. Select: the perpendicular line passing through point C, line $\overleftrightarrow{AB}$, [CONSTRUCT], [INTERSECTION] and deselect.
8. Select: line $\overleftrightarrow{CD}$, [DISPLAY] and [HIDE PERPENDICULAR LINE].
9. Select: points A, then C, then B, [CONSTRUCT], [SEGMENTS], [DISPLAY], [LINE STYLE], [MEDIUM], [DISPLAY], [LINE STYLE], [SOLID] and deselect.
10. Select: points C, D, [CONSTRUCT], [SEGMENT], [DISPLAY], [COLOR], click on the color you like and deselect.
11. Select: points A, then D, then C, [MEASURE], [ANGLE] and deselect.
12. Select: points B, then D, then C, [MEASURE], [ANGLE] and deselect.
13. Select: the Text tool, double click in the blank region to open a dialog box and (1) describe your observations about the measures of $\angle ADC$ and $\angle BDC$, (2) explain if altitude $\overline{CD}$ is perpendicular to side $\overline{AB}$. Justify your answer with the angle measures of $\angle ADC$ and $\angle BDC$.

14. Select: the Arrow tool, click on point C and, keeping the left button depressed, drag it first to the left of points A and B and then drag it to the right of points A and B. Observe the values that remain equal as you do so.
15. Select: the Text tool, double click in the blank region to open a dialog box and explain what you learned about the altitude of the triangle. Give your own definition of an altitude of a triangle.

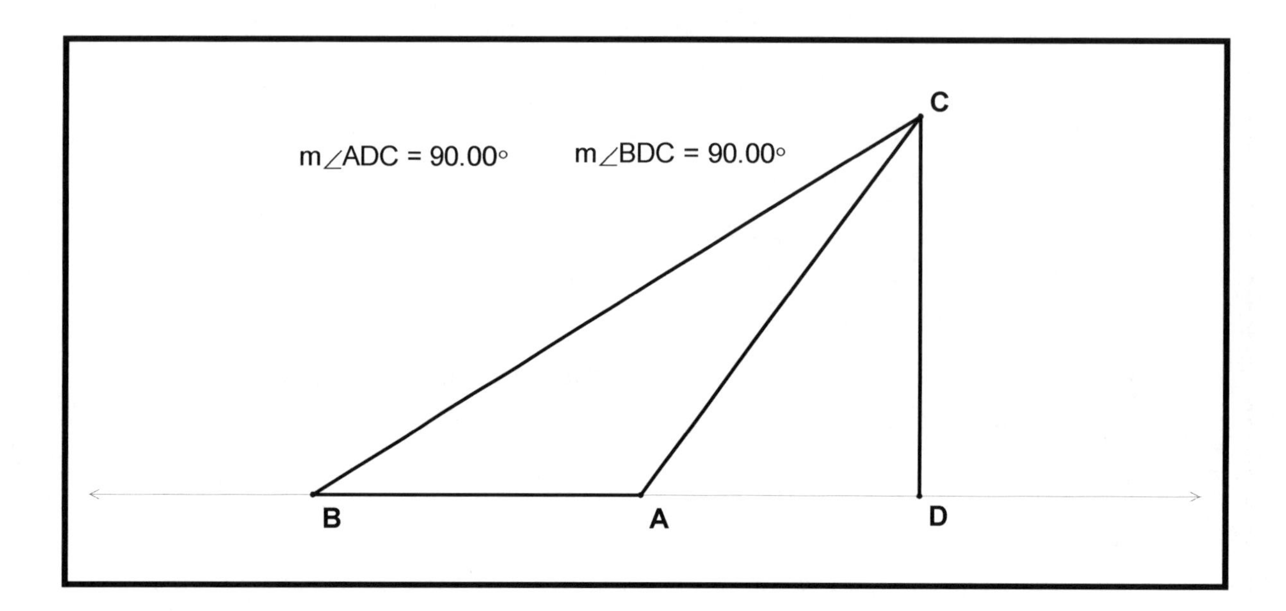

SUGGESTED EXERCISES

Type the solutions to the following problems in the blank region of your sketch.

1. $\overline{CD}$ is the altitude of ΔABC that is drawn from vertex C to side $\overline{AB}$. Find m∠ADC if m∠BDC = 90°.

2. $\overline{CD}$ is the altitude of ΔABC drawn from vertex C to side $\overline{AB}$. Find the value of x if m∠ADC = 2x and m∠BDC = 90°.

3. $\overline{AD}$ is the altitude of ΔABC drawn from vertex A to side $\overline{BC}$. Find the value of x if m∠ADB = 4x – 10 and m∠ADC = 90°.

LAB #8: THE MEDIANS OF A TRIANGLE

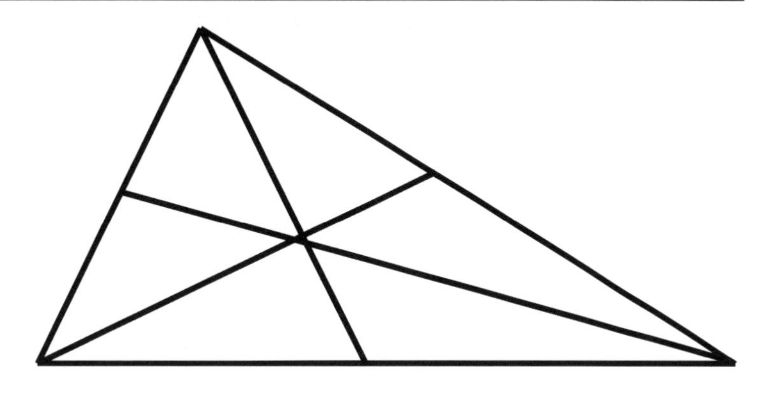

1. Select: [EDIT], [PREFERENCES], [TEXT], check the box [FOR ALL NEW POINTS] and click [OK].
2. Select: the Point tool and plot three points randomly as the vertices of a triangle.
3. Select: the Arrow tool, [EDIT], [SELECT ALL], [CONSTRUCT], [SEGMENTS] and deselect.
4. Select: side $\overline{AC}$, [CONSTRUCT], [MIDPOINT] and deselect.
5. Select: side $\overline{AB}$, [CONSTRUCT], [MIDPOINT] and deselect.
6. Select: side $\overline{BC}$, [CONSTRUCT], [MIDPOINT] and deselect.
7. Select: point B, midpoint D, [CONSTRUCT], [SEGMENT], [DISPLAY], [COLOR], click on the color you like, [OK] and deselect.
8. Select: point C, midpoint E, [CONSTRUCT], [SEGMENT] and deselect.
9. Select: point A, midpoint F, [CONSTRUCT], [SEGMENT] and deselect.
10. Select: points A, D, [MEASURE], [DISTANCE] and deselect.
11. Do the same to measure line segment $\overline{CD}$.
12. Select: the Text tool, double click in the blank region to open a dialog box and describe your observations about the measures of line segments $\overline{AD}$ and $\overline{CD}$.
13. Select: the Arrow tool and click in the blank region to deselect.
14. Select: points A, E, [MEASURE], [DISTANCE] and deselect.
15. Do the same to measure line segment $\overline{BE}$.

16. Select: the Text tool, double click in the blank region to open the dialog box and describe your observations about the measures of line segments $\overline{AE}$ and $\overline{BE}$.
17. Select: the Arrow tool and click in the blank region to deselect.
18. Select: points B, F, [MEASURE], [DISTANCE] and deselect.
19. Do the same to measure line segment $\overline{CF}$.
20. Select: the Text tool, double click in the blank region to open a dialog box and describe your observations about the measures of line segments $\overline{BF}$ and $\overline{CF}$.
21. Select: the Arrow tool, click on any labeled point and keeping the left button depressed, drag it. Observe the values that remain equal as you do so.
22. Select: the Text tool, double click in the blank region to open a dialog box and explain what the medians $\overline{AF}$, $\overline{BD}$, and $\overline{CE}$ do to sides $\overline{BC}$, $\overline{AC}$, and $\overline{AB}$, respectively, of ΔABC. Give your own definition of the median of the triangle.

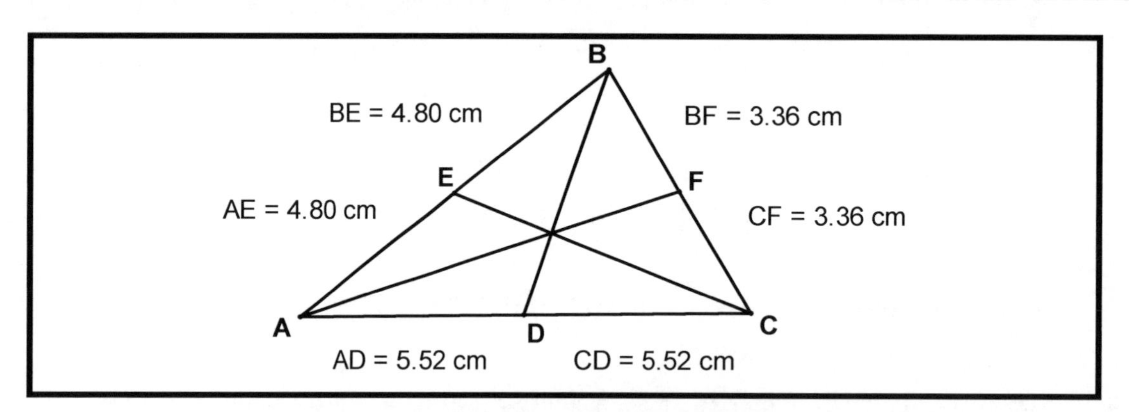

SUGGESTED EXERCISES

Type the solutions to the following problems in the blank region of your sketch.

1. $\overline{CE}$ is the median of ΔABC drawn from vertex C to side $\overline{AB}$.
   If $m\overline{AE}$ = 3 in, find the value of $\overline{BE}$.

2. $\overline{BD}$ is the median of ΔABC drawn from vertex B to side AC.
   If $m\overline{AD} = 3x$ and $m\overline{CD}$ = 21 cm, find the value of x.

3. $\overline{AF}$ is the median of ΔABC drawn from vertex A to side BC.
   If $m\overline{BF} = 4x + 20$ and $m\overline{CF} = 2x + 30$, find the value of x and measures of $\overline{BF}$, $\overline{CF}$, and $\overline{BC}$.

LAB #9: COMPLEMENTARY ANGLES

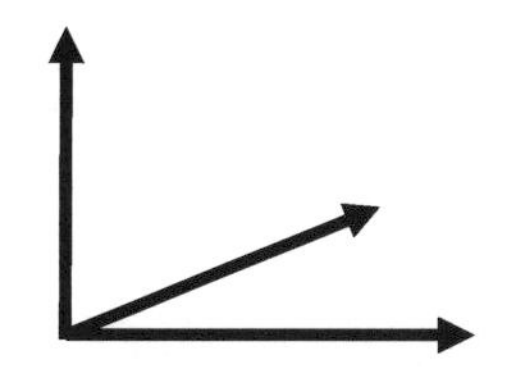

1. Select: [EDIT], [PREFERENCES], [TEXT], check the box [FOR ALL NEW POINTS] and click [OK].
2. Select: the Line Segment tool and construct a horizontal line segment in the direction from left to right approximately one inch long in the middle of your screen.
3. Select: the Arrow tool and click in the blank region to deselect.
4. Select: points A, then B, [CONSTRUCT], [CIRCLE BY CENTER + POINT] and deselect.
5. Select: point A, line segment $\overline{AB}$, [CONSTRUCT], [PERPENDICULAR LINE] and deselect.
6. Select: the circle, perpendicular line passing through point A, [CONSTRUCT], [INTERSECTIONS] and deselect.
7. Select: point D, line $\overleftrightarrow{CD}$, [DISPLAY], and [HIDE OBJECTS].
8. Select: points A, C, [CONSTRUCT], [SEGMENT] and deselect.
9. Select: points B, then point C, the circle, [CONSTRUCT], [ARC ON CIRCLE], [CONSTRUCT], [POINT ON ARC], [EDIT], [SELECT PARENTS], [DISPLAY] and [HIDE ARC].
10. Select: points A, E, [CONSTRUCT], [SEGMENT] and deselect.
11. Select: the dashed circle, [DISPLAY] and [HIDE CIRCLE].
12. Select: points B, then A, then C, [MEASURE], [ANGLE] and deselect.
13. Do the same to measure $\angle$CAE and $\angle$BAE.
14. Select: [NUMBER], [CALCULATE], click on the caption that shows m$\angle$CAE, [+], click on the caption that shows m$\angle$BAE and [OK].
15. Select: the Text tool, double click in the blank region to open a

dialog box, and describe your observations about the measure of ∠BAC and the value of m∠CAE + m∠BAE.

16. Select: the Arrow tool and click in the blank region to deselect.
17. Select: point E, [EDIT], [ACTION BUTTONS], [ANIMATION], [MEDIUM], [SLOW], [OK] and deselect.
18. Click on [ANIMATE POINT] and observe the measure of ∠BAC and the value of m∠CAE + m∠BAE.
19. Click on [ANIMATE POINT] again to stop the animation.
20. Select: the Text tool, double click in the blank region to open a dialog box and (1) explain to what angle measure the sum of complementary angles ∠CAE and ∠BAE is always equal, (2) write your own definition of complementary angles.

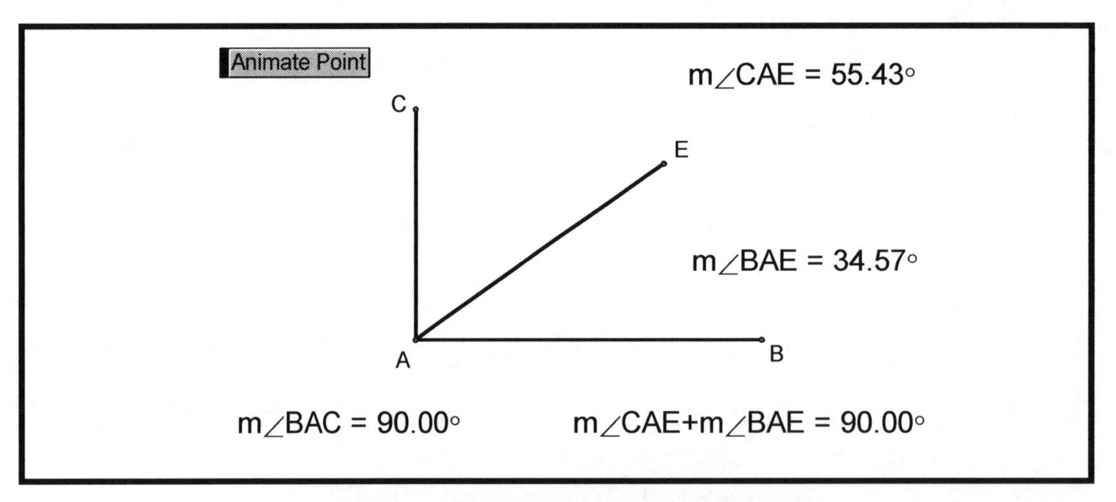

SUGGESTED EXERCISES

Type the solutions to the following problems in the blank region of your sketch.

1. ∠BAC is a right angle. m∠CAE = 37°. If ∠BAE is complementary to ∠CAE, find m∠ BAE.

2. ∠CAE and ∠BAE are complementary. If m∠CAE = 2x + 10, and m∠BAE = x + 20, find the value of x, m∠CAE and m∠BAE.

3. Two angles are complementary. One angle is twice as large as the other. Find the number of degrees in each angle.

LAB #10: SUPPLEMENTARY ANGLES

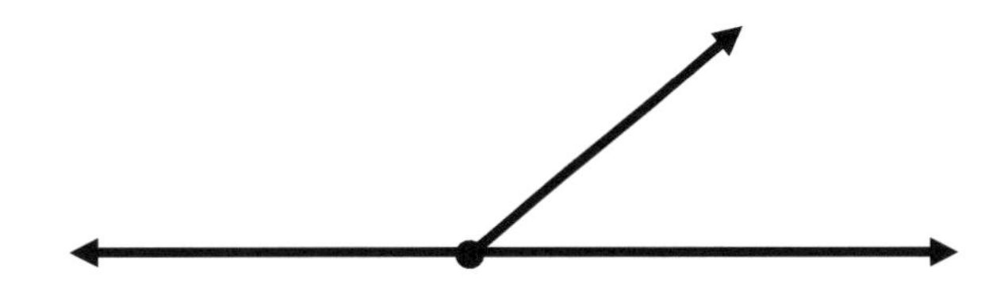

1. Select: [EDIT], [PREFERENCES], [TEXT], check the box [FOR ALL NEW POINTS] and click [OK].
2. Select: the Point tool and plot two points approximately two inches apart from each other in the middle of your screen.
3. Select: the Arrow tool, [EDIT], [SELECT ALL], [CONSTRUCT], [SEGMENT], [CONSTRUCT], [MIDPOINT] and deselect.
4. Select: points C, then B, [CONSTRUCT], [CIRCLE BY CENTER + POINT], [CONSTRUCT], [POINT ON CIRCLE], [EDIT], [SELECT PARENTS], [DISPLAY] and [HIDE CIRCLE].
5. Select: point C, D, [CONSTRUCT], [SEGMENT] and deselect.
6. Select: points A, then C, then B, [MEASURE], [ANGLE] and deselect.
7. Do the same to measure $\angle$ACD and $\angle$BCD.
8. Select: [NUMBER], [CALCULATE], click on the caption that shows m$\angle$ACD, [+], click on the caption that shows m$\angle$BCD and [OK].
9. Select: the Text tool, double click in the blank region to open a dialog box and describe your observations about the measure of $\angle$ACB and the value of m$\angle$ACD + m$\angle$BCD.
10. Select: the Arrow tool and click in the blank region of your sketch to deselect.
11. Select: point D, [EDIT], [ACTION BUTTONS], [ANIMATION], [MEDIUM], [SLOW] and [OK].
12. Click on [ANIMATE POINT] and observe the values that remain constant.

13. Click on [ANIMATE POINT] again to stop the animation.

14. Select: the Text tool, double click in the blank region to open a dialog box, and (1) explain to what angle measure is the sum of supplementary angles ∠ACD and ∠BCD always equal, (2) write your own definition of supplementary angles.

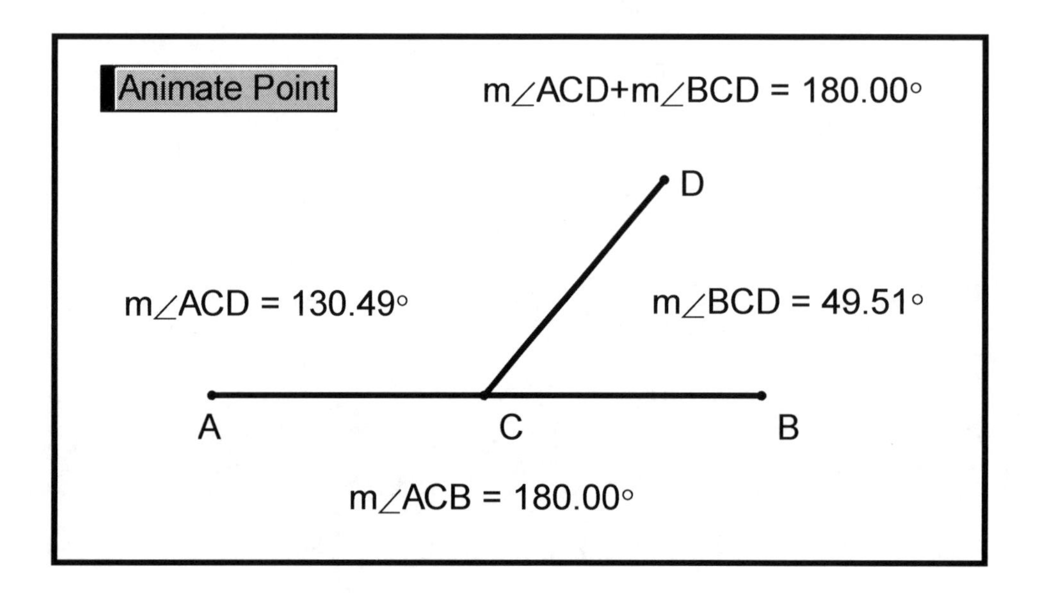

SUGGESTED EXERCISES

Type the solutions to the following problems in the blank region of your sketch.

1. Two angles are supplementary. Find the measure of the second angle if the measure of the first angle is 69°.

2. ∠ ACD and ∠BCD are supplementary. If m∠ACD = x and m∠BCD = 3x, find the value of x and the measure of each angle.

3. ∠A and ∠B are supplementary. If m∠A = x + 30 and m∠B = 3x - 10, find the value of x and the measure of each angle.

4. The measure of the supplement of an angle is 40° more than the measure of the angle. Find the number of degrees in the supplement of the angle.

LAB #11: VERTICAL ANGLES

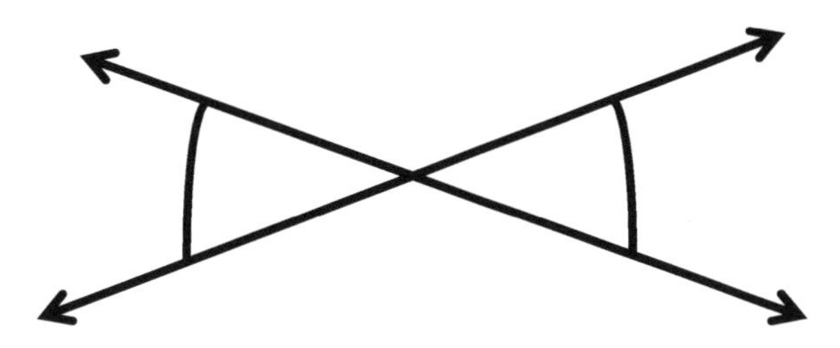

1. Select: [EDIT], [PREFERENCES], [TEXT], check the box [FOR ALL NEW POINTS] and click [OK].
2. Select: the Point tool and plot two points approximately two inches apart from each other in the middle of your screen.
3. Select: the Arrow tool, [EDIT], [SELECT ALL], [CONSTRUCT], [SEGMENT], [CONSTRUCT], [MIDPOINT] and deselect.
4. Select: points C, then B, [CONSTRUCT], [CIRCLE BY CENTER + POINT], [CONSTRUCT], [POINT ON CIRCLE] and deselect.
5. Select: point C, [TRANSFORM] and [MARK CENTER].
6. Select: point D, [TRANSFORM], [ROTATE], type 180 in the displayed window marked as [DEGREES] and click [ROTATE]. **KEEP THE NEW POINT SELECTED!**
7. Select: [EDIT], [PROPERTIES], [LABEL], type a capital E in the displayed window, click [OK] and deselect.
8. Select: points D, E, [CONSTRUCT], [SEGMENT] and deselect.
9. Select: the circle, [DISPLAY] and [HIDE CIRCLE].
10. Select: points A, then C, then D, [MEASURE], [ANGLE] and deselect.
11. Do the same to measure $\angle$BCE, $\angle$ACE, and $\angle$BCD.
12. Select: the Text tool, double click in the blank region to open a dialog box and (1) describe your observations about the measures of vertical angles $\angle$ACD and $\angle$BCE,
    (2) describe your observations about the measures of vertical angles $\angle$ACE and $\angle$BCD.

13. Select: the Arrow tool and click in the blank region to deselect.

    Select: point D, [EDIT], [ACTION BUTTONS], [ANIMATION], [MEDIUM], [OTHER], type [0.2], click [OK] and deselect.

14. Click on [ANIMATE POINT] and observe the angles that remain equal.
15. Click on [ANIMATE POINT] again to stop the animation.
16. Select: the Text tool, double click in the blank region to open a dialog box and (1) explain what you learned about the first pair of vertical angles ∠ACD and ∠BCE, (2) explain what you learned about the second pair of vertical angles ∠ACE and ∠BCD, (3) give your own definition of vertical angles.

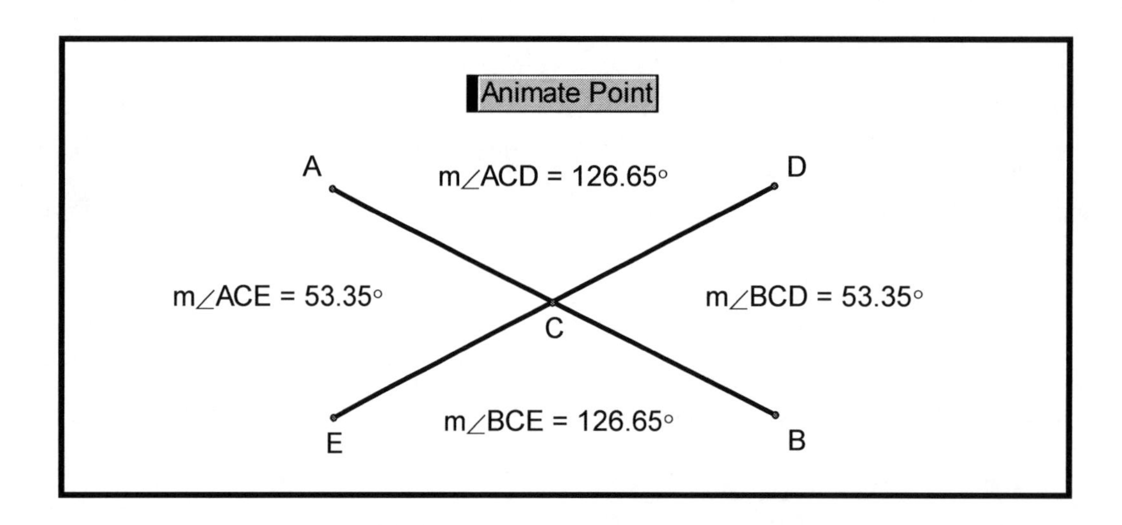

SUGGESTED EXERCISES

Type the solutions to the following problems in the blank region of your sketch.

1. ∠BCD and ∠ACE are vertical angles. Find the measure of m∠BCD if m∠ACE = 61°.

2. Line segments $\overline{AB}$ and $\overline{DE}$ intersect at C. If m∠ ACE = 5x and m∠BCD = 3x + 10, find x and m∠ACE.

3. Line segments $\overline{MN}$ and $\overline{RS}$ intersect at T. If m∠RTM = 7x + 16 and m∠NTS = 3x + 48, find x and m∠NTS.

LAB #12: PARALLEL LINES – THE CORRESPONDING ANGLES

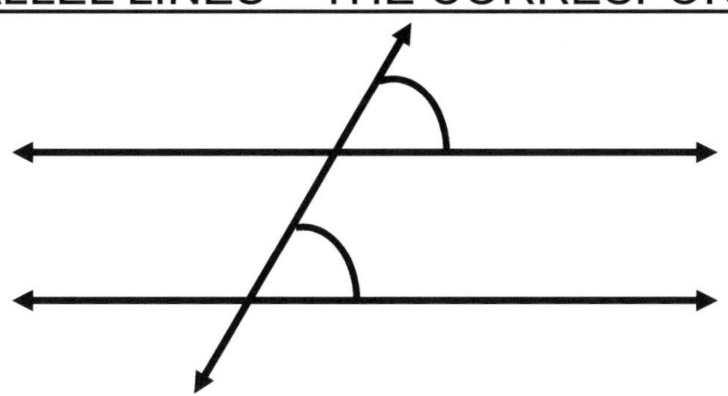

1. Select: [EDIT], [PREFERENCES], [TEXT], check the box [FOR ALL NEW POINTS] and click [OK].
2. Select: the Point tool and plot two points from left to right approximately three inches apart from each other slightly below the middle of your screen.
3. Select: the Arrow tool, [EDIT], [SELECT ALL], [CONSTRUCT], [LINE], [EDIT], [SELECT PARENTS], [CONSTRUCT], [SEGMENT], [CONSTRUCT], [MIDPOINT] and deselect.
4. Select: the Point tool and plot a point approximately one inch directly above point A.
5. Select: the Arrow tool and click in the blank region to deselect.
6. Select: point D, line segment $\overline{AB}$, [CONSTRUCT], [PARALLEL LINE], [CONSTRUCT], [POINT ON PARALLEL LINE] and deselect.
7. Click on point E, and keeping the left button depressed, drag it directly above point B and deselect.
8. Select: points D, E, [CONSTRUCT], [SEGMENT], [CONSTRUCT], [POINT ON SEGMENT], [EDIT], [ACTION BUTTONS], [ANIMATION], [MEDIUM], [SLOW], [OK] and deselect.
9. Select: points C, F, [CONSTRUCT], [LINE], [DISPLAY], [COLOR], click on the red bar, [CONSTRUCT], [POINT ON LINE], [EDIT], [SELECT PARENTS], [CONSTRUCT], [POINT ON LINE] and deselect.
10. Select: dashed lines $\overleftrightarrow{AB}$ and $\overleftrightarrow{DE}$, [DISPLAY], [LINE STYLE], [SOLID] and deselect.
11. Click on point G, and keeping the left button depressed, drag it approximately an inch above point F.

12. Click on point H, and keeping the left button depressed, drag it approximately an inch below point C.
13. Select: points G, then F, then E, [MEASURE], [ANGLE] and deselect.
14. Select: points F, then C, then B, [MEASURE], [ANGLE] and deselect.
15. Select: the Text tool, double click in the blank region to open a dialog box and write your observations about the measures of the corresponding angles $\angle GFE$ and $\angle FCB$.
16. Select: the Arrow tool and click in the blank region to deselect.
17. Click on [ANIMATE POINT] and observe what remains equal as point F travels along line segment $\overline{DE}$.
18. Click on [ANIMATE POINT] again to stop the animation.
19. Find three more pairs of corresponding angles in the diagram and measure them.
20. Select: the Text tool, double click in the blank region to open a dialog box and explain what you learned about the corresponding angles created by a transversal crossing two parallel lines.

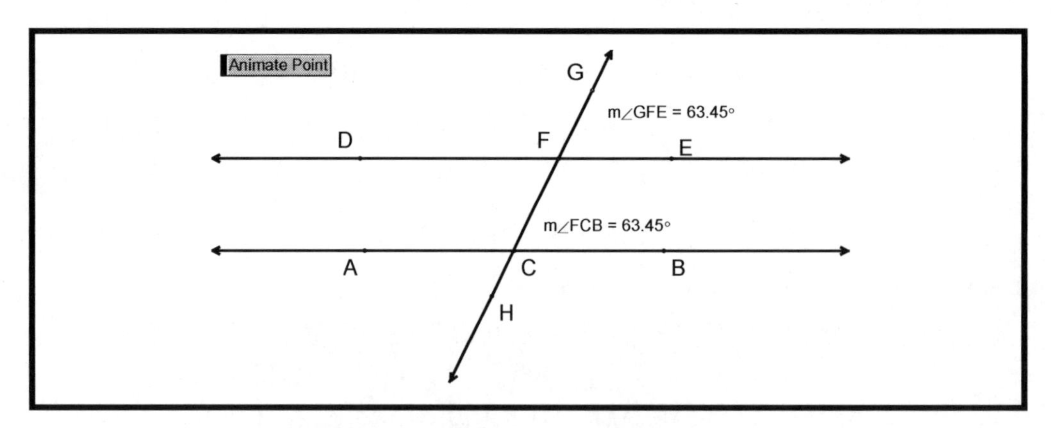

SUGGESTED EXERCISES

Type the solutions to the following problems in the blank region of your sketch.

1. In the diagram above, if $m\angle GFE = 73°$, find $m\angle FCB$.
2. $\angle HCB$ and $\angle CFE$ are corresponding angles. If $m\angle HCB = 4x$ and $m\angle CFE = 120°$, find the value of x.
3. $\angle GFE$ and $\angle FCB$ are corresponding angles. If $m\angle GFE = 2x + 40$ and $m\angle FCB = 3x + 20$, find the value of x and the measure of each angle.

LAB #13: PARALLEL LINES – THE ALTERNATE INTERIOR ANGLES

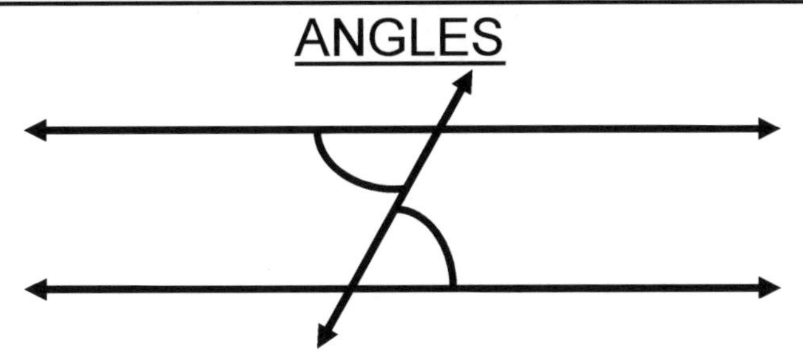

1. Select: [EDIT], [PREFERENCES], [TEXT], check the box [FOR ALL NEW POINTS] and click [OK].
2. Select: the Point tool and plot two points from left to right approximately three inches apart from each other slightly below the middle of your screen.
3. Select: the Arrow tool, [EDIT], [SELECT ALL], [CONSTRUCT], [LINE], [EDIT], [SELECT PARENTS], [CONSTRUCT], [SEGMENT], [CONSTRUCT], [MIDPOINT] and deselect.
4. Select: the Point tool and plot a point approximately one inch directly above point A.
5. Select: the Arrow tool and click in the blank region to deselect.
6. Select: point D, line segment $\overline{AB}$, [CONSTRUCT], [PARALLEL LINE], [CONSTRUCT], [POINT ON PARALLEL LINE] and deselect.
7. Click on point E, and keeping the left button depressed, drag it directly above point B and deselect.
8. Select: points D, E, [CONSTRUCT], [SEGMENT], [CONSTRUCT], [POINT ON SEGMENT], [EDIT], [ACTION BUTTONS], [ANIMATION], [MEDIUM], [SLOW], [OK] and deselect.
9. Select: points C, F, [CONSTRUCT], [LINE], [DISPLAY], [COLOR], click on the red bar, [CONSTRUCT], [POINT ON LINE], [EDIT], [SELECT PARENTS], [CONSTRUCT], [POINT ON LINE] and deselect.
10. Click on point G, and keeping the left button depressed, drag it approximately an inch above point F.
11. Click on point H, and keeping the left button depressed, drag it approximately an inch below point C.
12. Select: dashed lines $\overleftrightarrow{AB}$ and $\overleftrightarrow{DE}$, [DISPLAY], [LINE STYLE], [SOLID] and deselect.

13. Select: points E, then F, then C, [MEASURE], [ANGLE] and deselect.
14. Select: points A, then C, then F, [MEASURE], [ANGLE] and deselect.
15. Select: the Text tool, double click in the blank region to open a dialog box and write your observations about the measures of the alternate interior angles ∠EFC and ∠ACF.
16. Click on [ANIMATE POINT] and observe what remains the equal as point F travels along line segment $\overline{DE}$.
17. Click on [ANIMATE POINT] again to stop the animation.
18. Find one more pair of alternate interior angles and measure them.
19. Select: the Text tool, double click in the blank region to open a dialog box and explain what you learned about alternate interior angles.

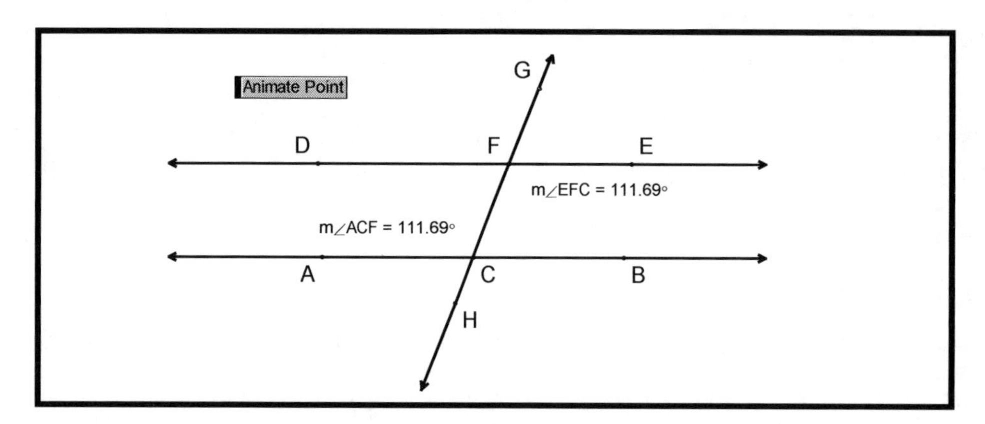

SUGGESTED EXERCISES

Type the solutions to the following problems in the blank region of your sketch.

1. $\overleftrightarrow{AB}$ // $\overleftrightarrow{DE}$. $\overleftrightarrow{CF}$ is a transversal that intersects each of these parallel lines at C and F, respectively. If m∠EFC = 78°, find m∠ACF.

2. $\overleftrightarrow{AB}$ // $\overleftrightarrow{DE}$. $\overleftrightarrow{CF}$ is a transversal that intersects each of these parallel lines at C and F, respectively. ∠DFC and ∠BCF are alternate interior angles. If m∠DFC = 135° and m∠BCF = 3x, find the value of x.

3. $\overleftrightarrow{AB}$ // $\overleftrightarrow{DE}$. $\overleftrightarrow{CF}$ is a transversal that intersects each of these parallel lines at C and F, respectively. ∠DFC and ∠BCF are alternate interior angles. If m∠DFC = 7x and m∠BCF = 3x + 60, find the value of x and the measure of each angle.

LAB #14: PARALLEL LINES – THE INTERIOR ANGLES ON THE SAME SIDE OF A TRANSVERSAL

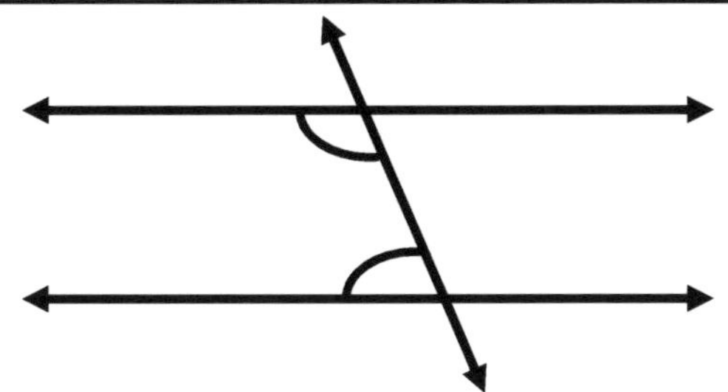

1. Select: [EDIT], [PREFERENCES], [TEXT], check the box [FOR ALL NEW POINTS] and click [OK].
2. Select: the Point tool and plot two points from left to right approximately three inches apart from each other slightly below the middle of your screen.
3. Select: the Arrow tool, [EDIT], [SELECT ALL], [CONSTRUCT], [LINE], [EDIT], [SELECT PARENTS], [CONSTRUCT], [SEGMENT], [CONSTRUCT], [MIDPOINT] and deselect.
4. Select: the Point tool and plot a point approximately one inch directly above point A.
5. Select: the Arrow tool and click in the blank region to deselect.
6. Select: point D, line segment $\overline{AB}$, [CONSTRUCT], [PARALLEL LINE], [CONSTRUCT], [POINT ON PARALLEL LINE] and deselect.
7. Click on point E, and keeping the left button depressed, drag it directly above point B and deselect.
8. Select: points D, E, [CONSTRUCT], [SEGMENT], [CONSTRUCT], [POINT ON SEGMENT], [EDIT], [ACTION BUTTONS], [ANIMATION], [MEDIUM], [SLOW], [OK] and deselect.
9. Select: points C, F, [CONSTRUCT], [LINE], [DISPLAY], [COLOR], click on the red bar, [CONSTRUCT], [POINT ON LINE], [EDIT], [SELECT PARENTS], [CONSTRUCT], [POINT ON LINE] and deselect.
10. Click on point G, and keeping the left button depressed, drag it approximately an inch above point F.
11. Click on point H, and keeping the left button depressed, drag it approximately an inch below point C.
12. Select: dashed lines $\overleftrightarrow{AB}$ and $\overleftrightarrow{DE}$, [DISPLAY], [LINE STYLE], [SOLID] and deselect.
13. Select: points E, then F, then C, [MEASURE], [ANGLE] and deselect.
14. Select: points B, then C, then F, [MEASURE], [ANGLE] and deselect.

15. Select: [NUMBER], [CALCULATE], click on the caption that shows m∠EFC, [+], click on the caption that shows m∠BCF, [OK] and deselect.
16. Select: the Text tool, double click in the blank region to open a dialog box, and (1) describe your observations about the measures of the interior angles on the same side of the transversal, (2) explain to what angle measure is their sum equal.
17. Select: the Arrow tool and click in the blank region to deselect.
18. Click on [ANIMATE POINT] and observe what remains the same as point F travels along line segment $\overline{DE}$.
19. Click on [ANIMATE POINT] again to stop the animation.
20. Find another pair of interior angles on the same side of the transversal and calculate their sum.
21. Select the Text tool, double click in the blank region to open a dialog box and explain what you learned about the interior angles on the same side of the transversal.

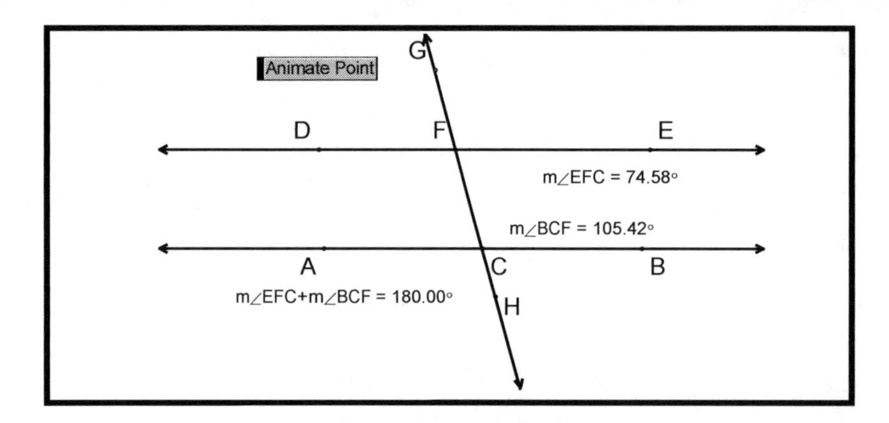

SUGGESTED EXERCISES

Type the solutions to the following problems in the blank region of your sketch.

1. In the diagram above, ∠EFC and ∠BCF are interior angles on the same side of the transversal. If m∠EFC = 78°, find m∠BCF.
2. $\overleftrightarrow{AB}$ // $\overleftrightarrow{DE}$. $\overleftrightarrow{CF}$ is a transversal that intersects each of these parallel lines at C and F, respectively. ∠EFC and ∠BCF are interior angles on the same side of the transversal. If m∠EFC = 3x and m∠BCF = 114°, find the value of x.
3. $\overleftrightarrow{AB}$ // $\overleftrightarrow{DE}$. $\overleftrightarrow{CF}$ is a transversal that intersects each of these parallel lines at C and F, respectively. If m∠ACF = 2x, and m∠DFC = 3x + 40, find the value of x and the measure of each angle.

LAB #15: THE EXTERIOR ANGLE OF A TRIANGLE

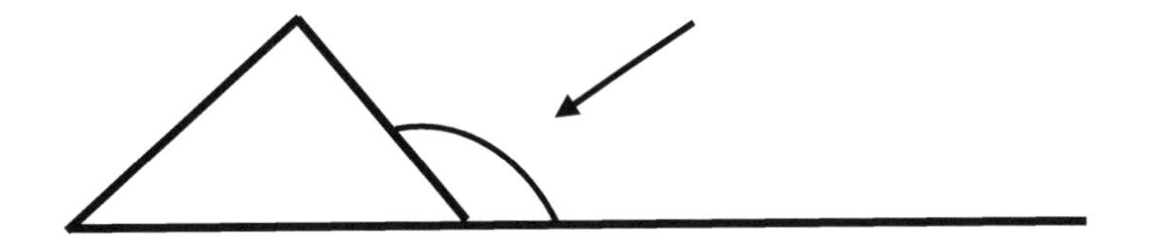

1. Select: [EDIT], [PREFERENCES], [TEXT], check the box [FOR ALL NEW POINTS] and click [OK].
2. Select: the Point tool and plot two points as the endpoints of a horizontal line segment in the middle of your screen.
3. Select: the Arrow tool, [EDIT], [SELECT ALL], [CONSTRUCT], [SEGMENT], [CONSTRUCT] and [POINT ON SEGMENT].
4. Click on point C and keeping the button depressed, drag it slightly closer to point B, and deselect.
5. Select: the Point tool and plot a point anywhere above line segment $\overline{AC}$.
6. Select: the Arrow tool and click in the blank region to deselect.
7. Select: points A, then C, then D, [CONSTRUCT], SEGMENTS] and deselect.
8. Select: points B, then C, then D, [MEASURE], [ANGLE], and deselect.
9. Do the same to measure the two interior angles $\angle$CAD and $\angle$ADC.
10. Select: [NUMBER], [CALCULATE], click on the caption that shows m$\angle$CAD, [+], click on the caption that shows m$\angle$ADC, [OK], and deselect.
11. Click on any labeled point, and keeping the left button depressed, drag it. Observe what values remain equal as you do so.

12. Select: dashed segment CB, [DISPLAY], [LINE STYLE], [SOLID] and deselect.
13. Select: the Text tool, double click in the blank region to open a dialog box, and explain what you have learned about the relationship between an exterior angle and the sum of two non-adjacent interior angles of the triangle.

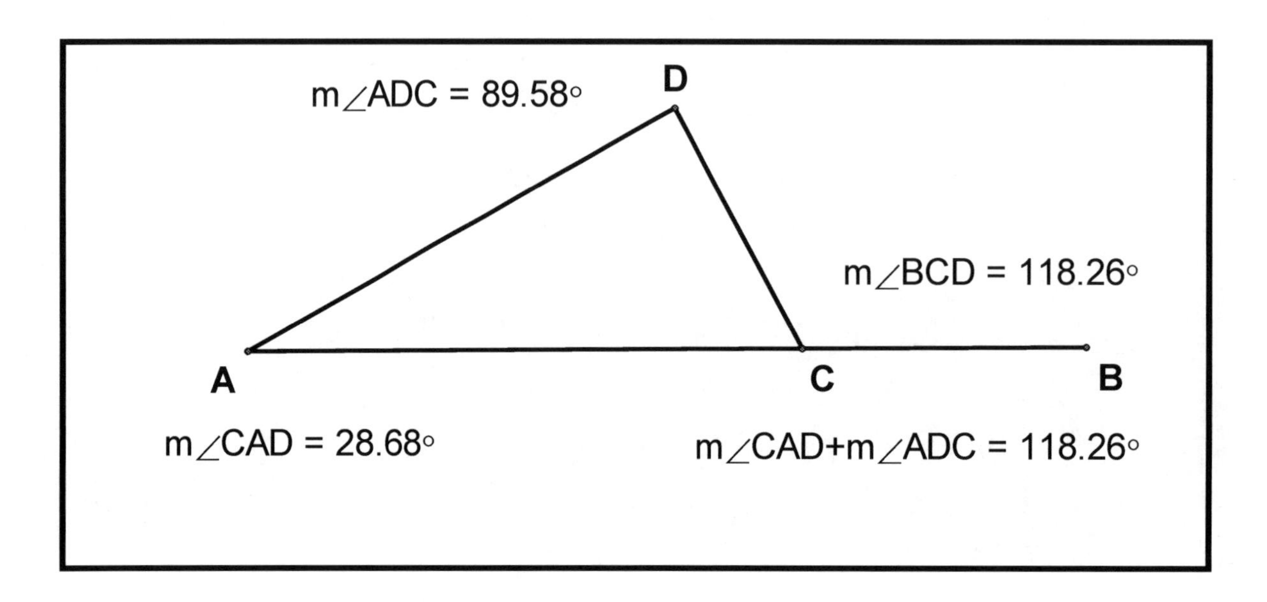

<u>SUGGESTED EXERCISES</u>

Type the solutions to the following problems in the blank region of your sketch.

1. In ΔADC, m$\angle$CAD = 50° and m$\angle$ADC = 85°. Find the degree measure of the exterior angle at vertex C.

2. In ΔADC , m$\angle$ADC = 90°, and the exterior angle at C is equal to 150°. Find the measure of $\angle$DAC.

3. In ΔADC, m$\angle$A = x and m$\angle$D= x +10. The measure of the exterior angle at C is 140°. Find the value of x and the measures of $\angle$A and $\angle$D.

LAB #16: THE SUM OF THE INTERIOR ANGLES OF A QUADRILATERAL

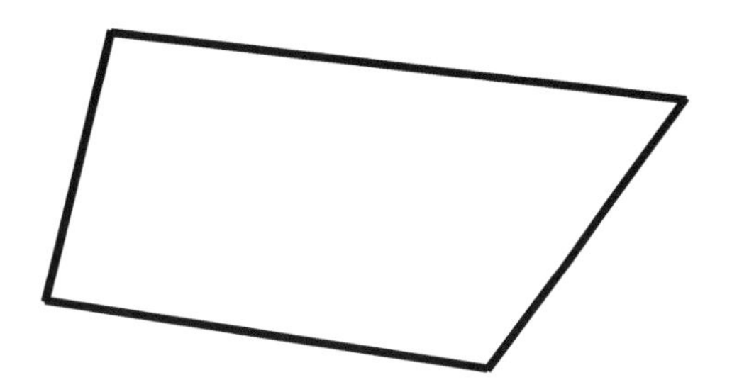

1. Select: [EDIT], [PREFERENCES], [TEXT], check the box [FOR ALL NEW POINTS] and click [OK].
2. Select: the Point tool and plot four points in a clockwise direction as the vertices of a quadrilateral.
3. Select: the Arrow tool, [EDIT], [SELECT ALL], [CONSTRUCT], [SEGMENTS], [CONSTRUCT], [POINTS ON SEGMENTS] and deselect.
4. Select: [EDIT], [SELECT ALL], [EDIT], [SELECT PARENTS], [DISPLAY] and [HIDE OBJECTS].
5. Select: [EDIT], [SELECT ALL], [CONSTRUCT], [SEGMENTS] and deselect.
6. Select: points E, then F, then G, [MEASURE], [ANGLE] and deselect.
7. Do the same to measure $\angle FGH$, $\angle GHE$, and $\angle HEF$.
8. Select: [NUMBER], [CALCULATE], click on the caption that shows $m\angle EFG$, [+], click on the caption that shows $m\angle FGH$, [+], click on the caption that shows $m\angle GHE$, [+], click on the caption that shows $m\angle HEF$, [OK] and deselect.
9. Select: points F, H, [EDIT], [ACTION BUTTONS], [ANIMATION], [MEDIUM], [SLOW] and [OK].
10. Click on [ANIMATE POINTS] and observe the value of the sum of the four interior angles of quadrilateral EFGH.

11. Click on [ANIMATE POINTS] again to stop the animation.
12. Select: the Text tool, double click in the blank region to open a dialog box and explain what you learned about the sum of the four interior angles of a quadrilateral.

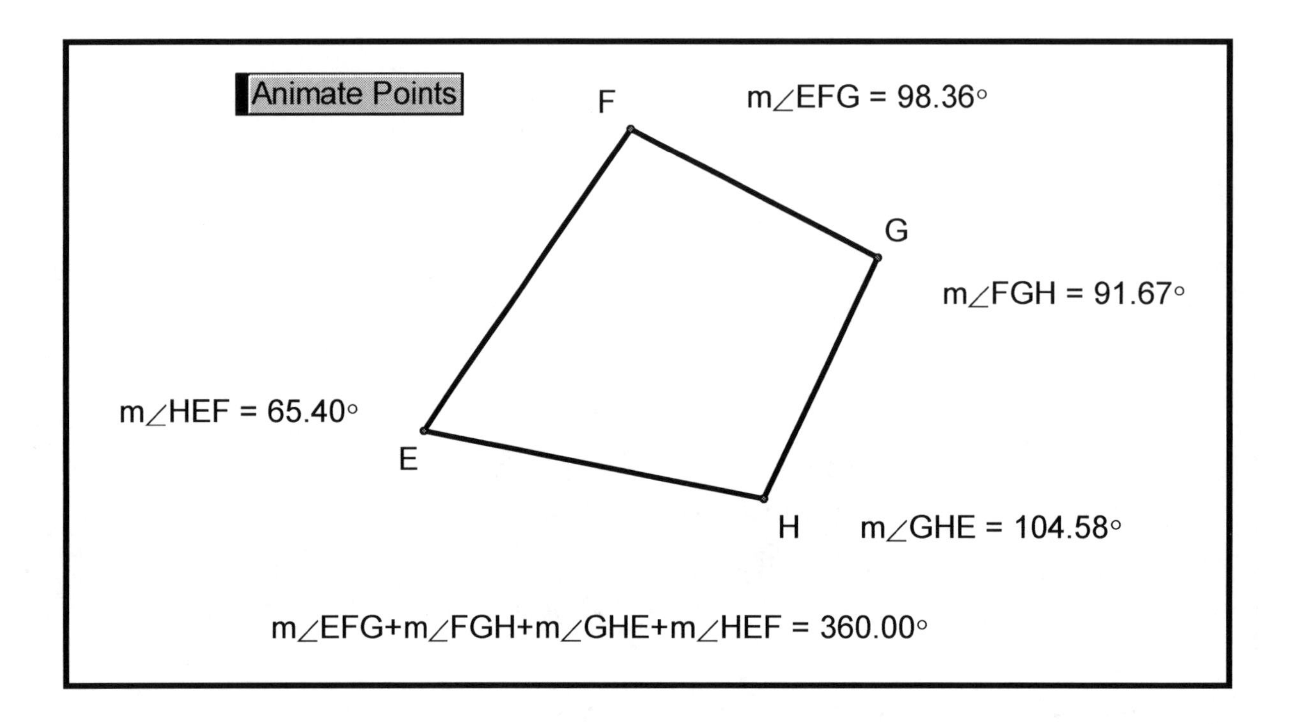

SUGGESTED EXERCISES

Type the solutions to the following problems in the blank region of your sketch.

1. EFGH is a quadrilateral. If $m\angle E = 73°$, $m\angle F = 64°$, and $m\angle G = 142°$. Find $m\angle H$.

2. EFGH is a quadrilateral. If $m\angle E = 81°$, $m\angle F = 70°$, $m\angle G = 139°$, and $m\angle H = 2x$ find the value of x.

3. EFGH is a quadrilateral. If $m\angle E = 46°$, $m\angle F = 93°$, $m\angle G = 2x + 6$, and $m\angle H = 3x – 5$. Find the value of x and the measure of the angles G and H.

LAB # 17: THE SUM OF THE INTERIOR ANGLES OF A POLYGON

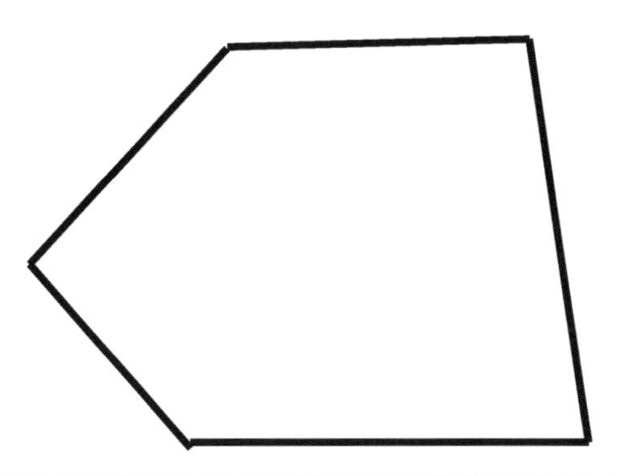

1. Select: [EDIT], [PREFERENCES], [TEXT], check the box [FOR ALL NEW POINTS] and click [OK].
2. Select: the Point tool and plot five points in a clockwise direction as the vertices of a pentagon.
3. Select: the Arrow tool, [EDIT], [SELECT ALL], [CONSTRUCT], [SEGMENTS], [CONSTRUCT], [POINTS ON SEGMENTS] and deselect.
4. Select: [EDIT], [SELECT ALL], [EDIT], [SELECT PARENTS], [DISPLAY] and [HIDE OBJECTS].
5. Select: [EDIT], [SELECT ALL], [CONSTRUCT], [SEGMENTS] and deselect.
6. Select: points F, then G, then H, [MEASURE], [ANGLE] and deselect.
7. Do the same to measure ∠GHI, ∠HIJ, ∠IJF, and ∠JFG.
8. Select: [NUMBER], [CALCULATE], click on the caption that shows m∠FGH, [+], click on the caption that shows m∠GHI, [+], click on the caption that shows m∠HIJ, [+], click on the caption that shows m∠IJF, [+], click on the caption that shows m∠JFG, [OK] and deselect.
9. Select: points F, H, I, [EDIT], [ACTION BUTTONS], [ANIMATION], [MEDIUM], [SLOW] and [OK].
10. Click on [ANIMATE POINTS] and observe the value of the sum of the five interior angles of quadrilateral FGHIJ.

11. Click on [ANIMATE POINTS] again to stop the animation.
12. Select: the Text tool, double click to open a dialog box and explain what you learned about the sum of the five interior angles of a pentagon.
13. Select: the Arrow tool, [NUMBER], [CALCULATE], 180, [∗], [( ], [5], [–], [2], [ )] and [OK] to evaluate the formula 180(n – 2) where n = 5 represents the number of sides of the constructed polygon
14. Select: the Text tool, double click in the blank region to open a dialog box, compare your answer with the sum of the interior angles of the constructed polygon and explain what you have to subtract and multiply to find the sum of interior angles of any polygon.

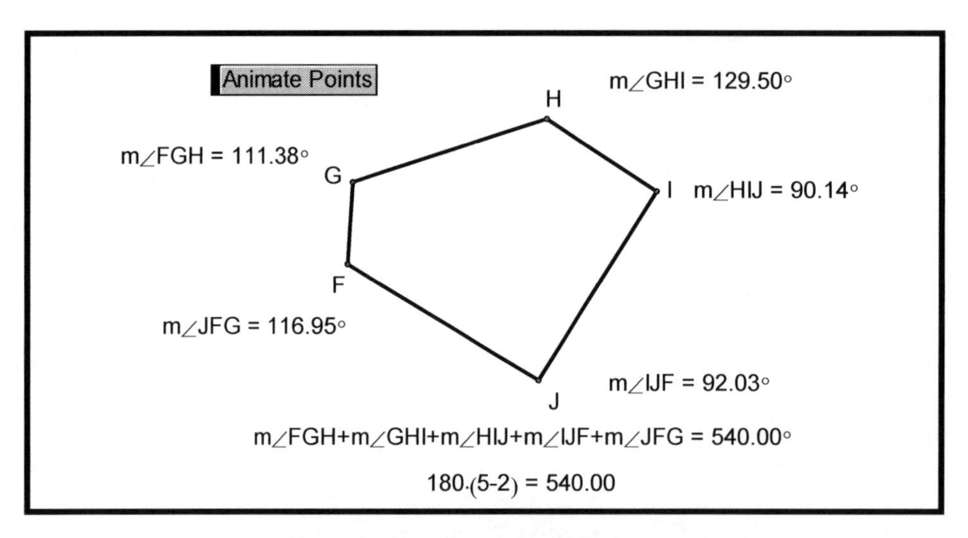

SUGGESTED EXERCISES

Type the solutions to the following problems in the blank region of your sketch.

1. Use the formula 180(n – 2) to find the sum of the angles of the polygon with 6 sides (a hexagon).
2. Use the formula 180(n – 2) to find the sum of the angles of the polygon with 12 sides (a dodecagon).
3. Use the formula 180(n – 2) to find the number of sides (n) of the polygon whose sum of the measures of the interior angles is equal to 2520°.

LAB #18: THE SUM OF THE EXTERIOR ANGLES OF A TRIANGLE

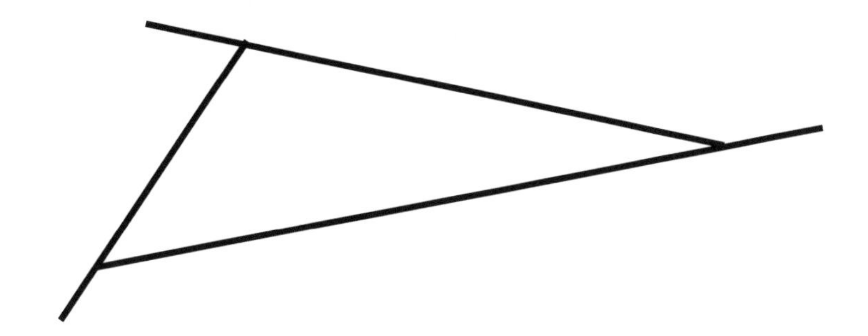

1. Select: [EDIT], [PREFERENCES], [TEXT], check the box [FOR ALL NEW POINTS] and click [OK].
2. Select: the Point tool and plot 3 points randomly as the vertices of a triangle.
3. Select: the Arrow tool, [EDIT], [SELECT ALL], [CONSTRUCT], [RAYS] and deselect.
4. Select: ray $\overrightarrow{CA}$, [CONSTRUCT] and [POINT ON RAY].
5. Click on point D, and keeping the left button depressed, drag it <u>outside</u> of ΔABC and deselect.
6. Select: ray $\overrightarrow{AB}$, [CONSTRUCT] and [POINT ON RAY].
7. Click on point E, and keeping the left button depressed, drag it <u>outside</u> of ΔABC and deselect.
8. Select: ray $\overrightarrow{BC}$, [CONSTRUCT] and [POINT ON RAY].
9. Click on point F, and keeping the left button depressed, drag it <u>outside</u> of ΔABC and deselect.
10. Select: points D, then A, then B, [MEASURE], [ANGLE] and deselect.
11. Do the same to measure the exterior angles $\angle EBC$ and $\angle FCA$.
12. Select: [NUMBER], [CALCULATE], click on the caption that shows $m\angle DAB$, [+], click on the caption that shows $m\angle EBC$, [+], click on the caption that shows $m\angle FCA$, [OK] and deselect.
13. Click on point C, and keeping the left button depressed, drag it. Observe what value remains constant as you do so.

14. Select: the Text tool, double click in the blank region to open a dialog box, and explain what you have learned about the sum of exterior angles of a triangle.

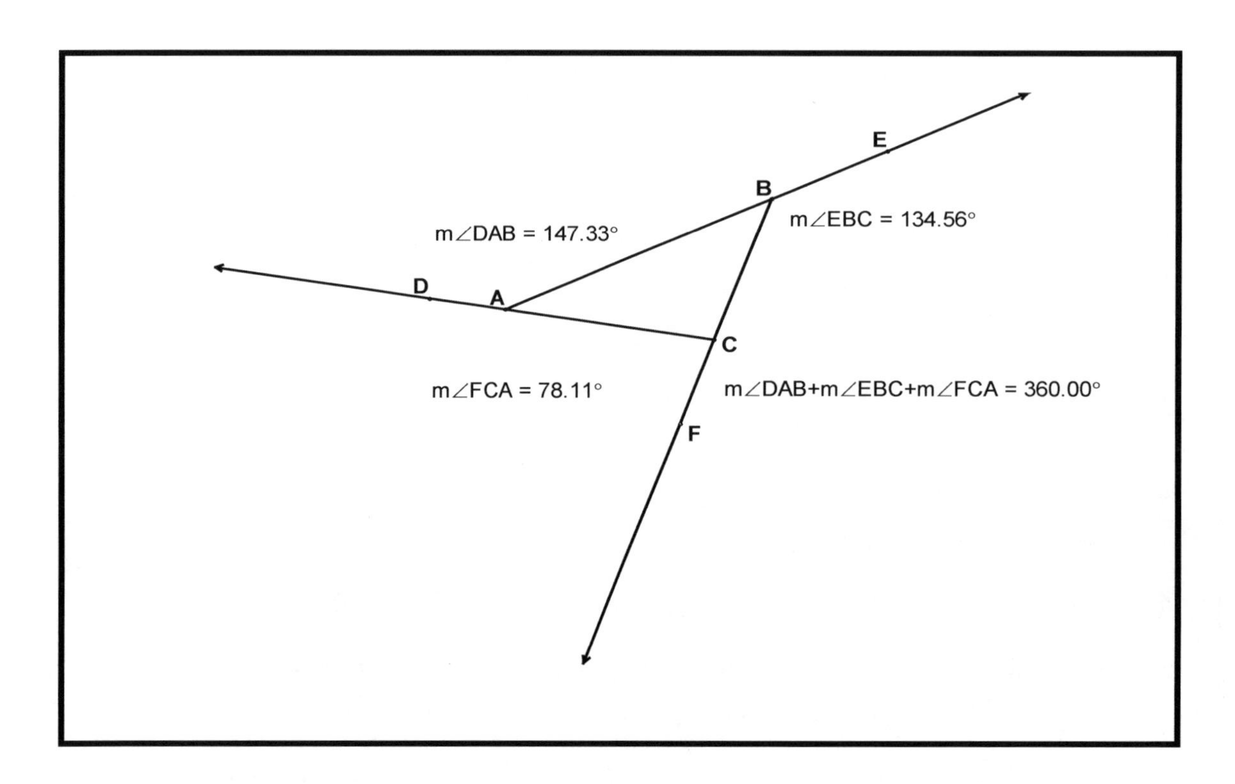

SUGGESTED EXERCISES

Type the solutions to the following problems in the blank region of your sketch.

1. ∠BAD, ∠ACF, and ∠CBE are the exterior angles of ΔABC.
   If m∠BAD = 112°, m∠ACF = 84°, find m∠CBE.

2. ∠BAD, ∠ACF, and ∠CBE are the exterior angles of ΔABC.
   If m∠BAD = 8x, m∠ACF = 120°, m∠CBE = 80°, find the value of x and m∠BAD.

3. ∠1, ∠2, and ∠3 are the exterior angles of a triangle.
   If m∠1 = x + 10, m∠2 = 3x – 20, and m∠3 = 2x + 40,
   find the value of x and the measure of each exterior angle.

LAB #19: THE SUM OF THE EXTERIOR ANGLES OF A QUADRILATERAL

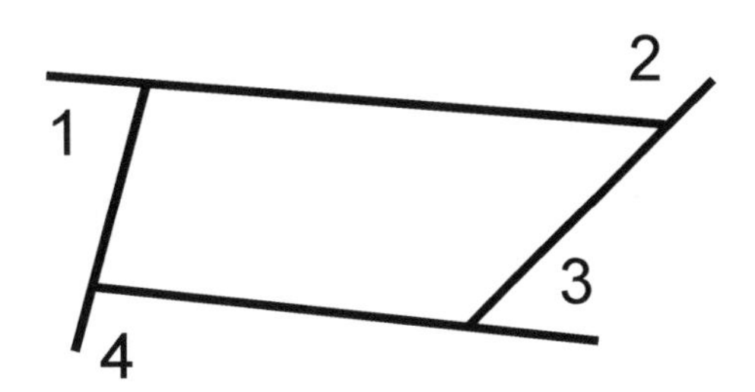

1. Select: [EDIT], [PREFERENCES], [TEXT], check the box [FOR ALL NEW POINTS] and click [OK].
2. Select: the Point tool and plot four points in a clockwise direction as the vertices of a quadrilateral.
3. Select: the Arrow tool, [EDIT], [SELECT ALL], [CONSTRUCT], [SEGMENTS], [CONSTRUCT], [POINTS ON SEGMENTS] and deselect.
4. Select: [EDIT], [SELECT ALL], [EDIT], [SELECT PARENTS], [DISPLAY] and [HIDE OBJECTS].
5. Select: [EDIT], [SELECT ALL], [CONSTRUCT], [RAYS], [CONSTRUCT], [POINTS ON RAYS] and deselect.
6. Click on the point(s) you see on the sides of the quadrilateral EFGH and, keeping the left button depressed, drag them <u>outside</u> of the quadrilateral EFGH and deselect.
7. Select: points L, then E, then F, [MEASURE], [ANGLE] and deselect.
8. Do the same to measure $\angle$IFG, $\angle$JGH, and $\angle$KHE.
9. Select: [NUMBER], [CALCULATE], click on the caption that shows m$\angle$LEF, [+], click on the caption that shows m$\angle$IFG, [+], click on the caption that shows m$\angle$JGH, [+], click on the caption that shows m$\angle$KHE, [OK] and deselect.
10. Select: points F, H, [EDIT], [ACTION BUTTONS], [ANIMATION], [MEDIUM], [SLOW] and [OK].
11. Click on [ANIMATE POINTS] and observe the value of the sum of the four exterior angles of quadrilateral EFGH.

12. Click on [ANIMATE POINTS] again to stop the animation.
13. Select: the Text tool, double click in the blank region to open a dialog box and explain what you have learned about the sum of exterior angles of a quadrilateral.

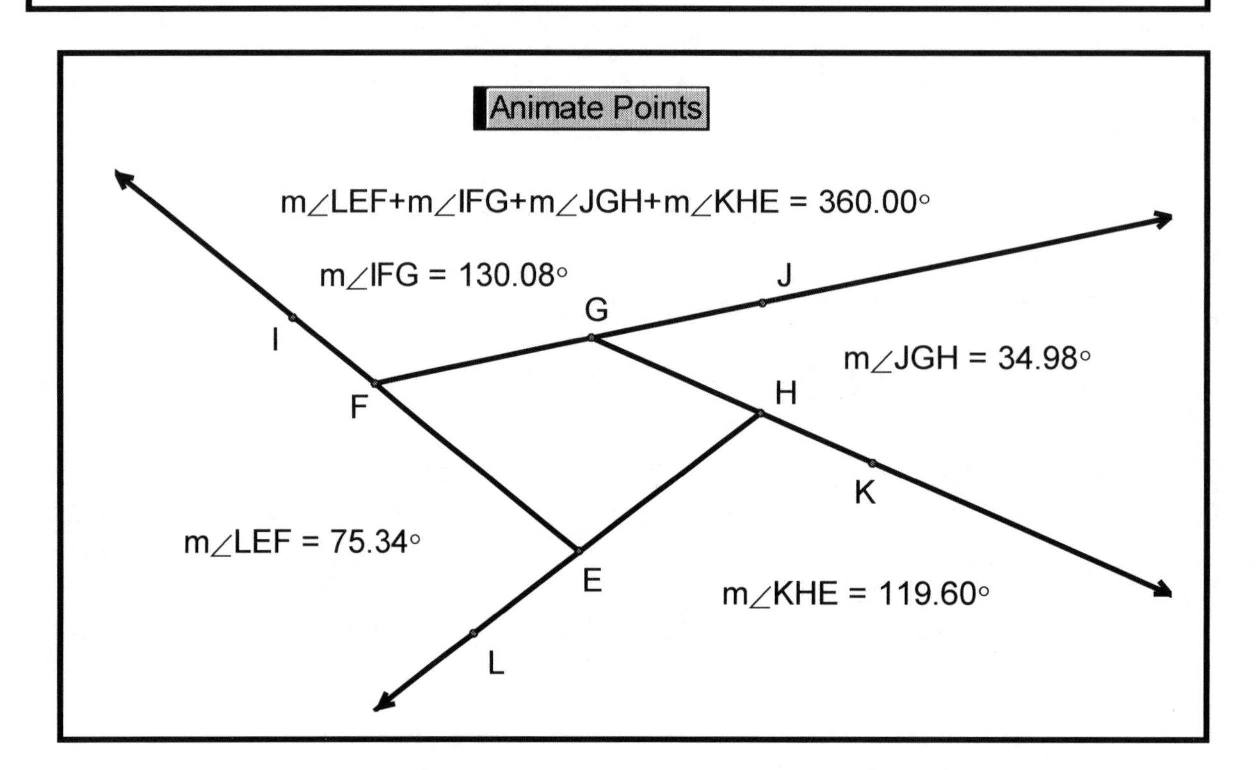

SUGGESTED EXERCISES

Type the solutions to the following problems in the blank region of your sketch.

1. $\angle 1$, $\angle 2$, $\angle 3$, and $\angle 4$ are the exterior angles of a quadrilateral. If $m\angle 1 = 101°$, $m\angle 2 = 112°$, and $m\angle 3 = 56°$, find $m\angle 4$.

2. $\angle LEF$, $\angle IFG$, $\angle JGH$, and $\angle KHE$ are the exterior angles of the quadrilateral ABCD. If $m\angle LEF = x$, $m\angle LFG = 2x$, $m\angle JGH = 4x$, and $m\angle KHE = 5x$, find the measure of each exterior angle.

3. $\angle 1$, $\angle 2$, $\angle 3$ and $\angle 4$ are the exterior angles of a quadrilateral. If $m\angle 1 = x + 10$, $m\angle 2 = 3x - 20$, $m\angle 3 = 2x + 40$ and $m\angle 4 = 5x$. Find the value of x and the measure of each of the exterior angles.

LAB #20: THE SUM OF THE EXTERIOR ANGLES OF A POLYGON

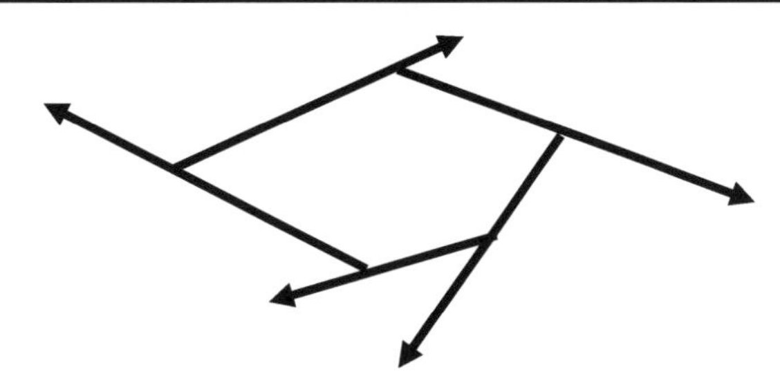

1. Select: [EDIT], [PREFERENCES], [TEXT], check the box [FOR ALL NEW POINTS] and click [OK].
2. Select: the Point tool and plot 5 points in a clockwise direction as the vertices of a pentagon.
3. Select: the Arrow tool, [EDIT], [SELECT ALL], [CONSTRUCT], [RAYS], [CONSTRUCT], [POINTS ON RAYS] and deselect.
4. Click on the point(s) you see on the sides of the pentagon ABCDE and, keeping the left button depressed, drag them <u>outside</u> of the pentagon ABCDE and deselect.
5. Select: points F, then A, then B, [MEASURE], [ANGLE] and deselect.
6. Do the same to measure the exterior angles ∠GBC, ∠HCD, ∠IDE and ∠JEA.
7. Select: [NUMBER], [CALCULATE], click on the caption that shows m∠FAB, [+], click on the caption that shows m∠GBC, [+], click on the caption that shows m∠HCD, [+], click on the caption that shows m∠IDE, [+], the caption that shows m∠JEA and [OK].
8. Select: the Text tool, double click in the blank region to open a dialog box and describe your observations about the sum of exterior angles of a pentagon.
9. Select: the Arrow tool and click in the blank region to deselect.
10. Select: points C, then D, [EDIT], [MERGE POINTS] and deselect
11. Select: points H, then D, then E, [MEASURE], [ANGLE] and deselect.
12. Select: [NUMBER], [CALCULATE], click on the caption that shows m∠FAB, [+], click on the caption that shows m ∠GBD, [+], click on the caption that shows m∠JEA, [+], click on the caption that shows m∠HDE and [OK].
13. Select: the Text tool, double click in the blank region to open a dialog box and describe your observations about the sum of exterior angles of a quadrilateral.
14. Select: the Arrow tool and click in the blank region to deselect.
15. Select: points D, then E, [EDIT], [MERGE POINTS] and deselect.
16. Select: points H, then E, then A, [MEASURE], [ANGLE] and deselect.

17. Select: [NUMBER], [CALCULATE], click on the caption that shows m∠FAB, [+], click on the caption that shows m∠GBE, [+], click on the caption that shows m∠HEA and [OK].
18. Select: the Text tool, double click in the blank region to open a dialog box and (1) describe your observations about the sum of exterior angles of a triangle,

    (2) explain what you have learned about the sum of exterior angles of any triangle, quadrilateral, pentagon or, in general, any polygon.

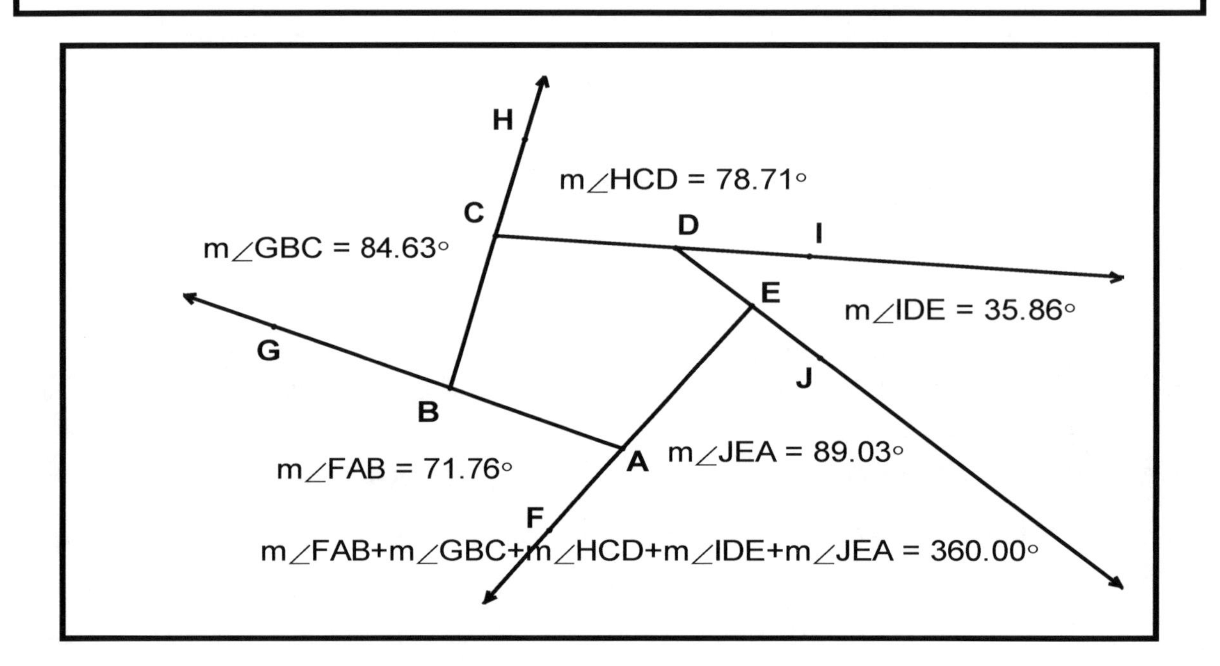

SUGGESTED EXERCISES

Type the solutions to the following problems in the blank region of your sketch.

1. ∠1, ∠2, ∠3, ∠4, and ∠5 are exterior angles of a pentagon. If m∠1 = 21°, m∠2 = 83°, m∠3 = 56°, m∠4 = 76°, find m∠5.

2. ∠1, ∠2, ∠3, and ∠4 are the exterior angles of quadrilateral ABCD. If m∠1 = x, m∠ 2 = 2x, m∠ 3 = 4x, and m∠4 = 5x, find the measure of each exterior angle.

3. ∠1, ∠2, and ∠3 are the exterior angles of a triangle. If m∠1 = x + 10, m∠2 = 3x – 20, and m∠3 = 2x + 40, find the value of x and each exterior angle.

LAB #21: PROPERTIES OF AN ISOSCELES TRIANGLE

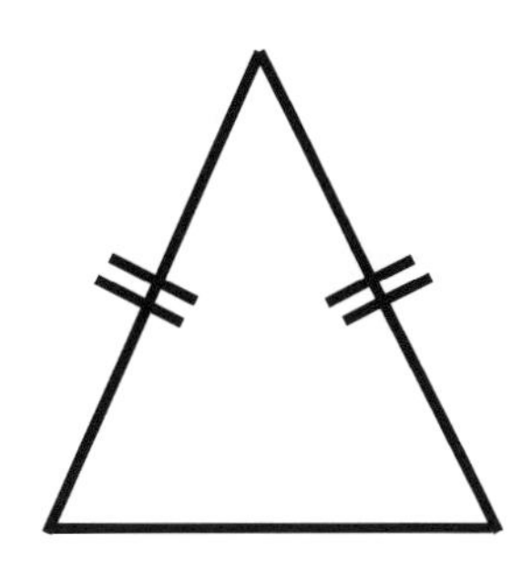

1. Select: [EDIT], [PREFERENCES], [TEXT], check the box [FOR ALL NEW POINTS] and click [OK].
2. Select: the Point tool and plot two points approximately one inch apart from each other in the middle of your screen.
3. Select: the Arrow tool and click in the blank region to deselect.
4. Select: points A, then B, [CONSTRUCT], [CIRCLE BY CENTER + POINT], [CONSTRUCT] and [POINT ON CIRCLE].
5. Click on point B, and keeping the left button depressed, drag it along the circle below and to the left of point A.
6. Click on point C, and keeping the left button depressed, drag it along the circle below and to the right of point A and deselect.
7. Select: the circle, [DISPLAY] and [HIDE CIRCLE].
8. Select: [EDIT], [SELECT ALL], [CONSTRUCT], [SEGMENTS], [MEASURE], [LENGTHS] and deselect.
9. Select: the Text tool, double click to open a dialog box and describe your observations about the measures of sides $\overline{AB}$ and $\overline{CA}$ of ΔABC.
10. Select: the Arrow tool and click in the blank region to deselect.
11. Select: points B, then A, then C, [MEASURE], [ANGLE] and deselect.
12. Do the same to measure $\angle ABC$ and $\angle ACB$.
13. Select: the Text tool, double click in the blank region to open a dialog box and describe your observations about the measures of the base angles $\angle ABC$ and $\angle ACB$ of ΔABC.

14. Select: the Arrow tool, click on point A or B, and, keeping the left button depressed, drag it. Observe the values that remain equal as you do so.
15. Select: the Text tool, double click in the blank region to open a dialog box and (1) explain what you learned about the angles and the sides of an isosceles triangle,
(2) give your own definition of an isosceles triangle.

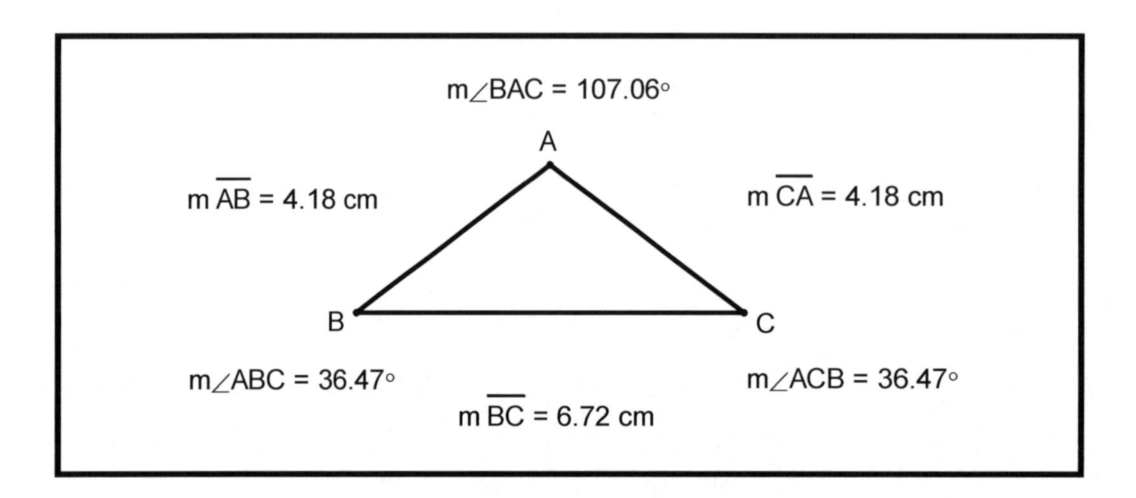

SUGGESTED PROBLEMS

Type the solutions to the following problems in the blank region of your sketch.

1. ΔABC is an isosceles triangle with congruent sides $\overline{AB}$ and $\overline{AC}$. Find the measure of side $\overline{AC}$ if side $m\overline{AB} = 7$ cm.

2. ΔABC is an isosceles triangle with congruent sides $\overline{AB}$ and $\overline{AC}$. Find the value of x if $m\overline{AB} = x + 6$ and $m\overline{AC} = 15$.

3. ΔABC is an isosceles triangle. $\overline{AB}$ and $\overline{AC}$ both equal 12 in. If $m\angle B = 70°$ and $m\angle C = 2x + 50$, find the value of x and $m\angle C$.

4. ΔABC is an isosceles triangle with $\overline{AB} \cong \overline{AC}$. If $m\angle ABC = 3x + 15$ and $m\angle ACB = x + 50$, find the value of x and the measure of each angle of ΔABC.

LAB #22: PROPERTIES OF THE EQUILATERAL TRIANGLE

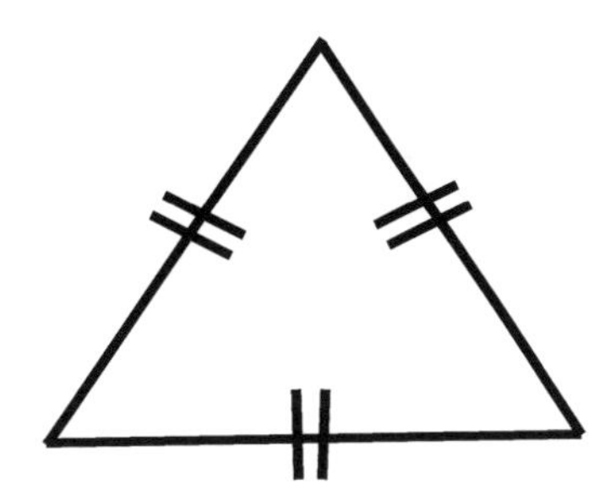

1. Select: [EDIT], [PREFERENCES], [TEXT], check the box [FOR ALL NEW POINTS] and click [OK].
2. Select: the Line Segment tool and construct a small horizontal line segment.
3. Select: the Arrow tool, [EDIT], [SELECT ALL], [CONSTRUCT], [CIRCLES BY CENTERS + RADIUS], [CONSTRUCT], [INTERSECTIONS] and deselect.
4. Select: the Arrow tool, both circles, point D, [DISPLAY] and [HIDE OBJECTS].
5. Select: points A, then C, then B, [CONSTRUCT], [SEGMENTS], [MEASURE], [LENGTHS] and deselect.
6. Select: the Text tool, double click in the blank region to open a dialog box and describe your observations about the measures of the sides of ΔABC.
7. Select: the Arrow tool and click in the blank region to deselect.
8. Select: points A, then B, then C, [MEASURE], [ANGLE] and deselect.
9. Do the same to measure $\angle ACB$ and $\angle BAC$.
10. Select: the Text tool, double click in the blank region to open a dialog box and describe your observations about the measures of the angles of ΔABC.
11. Select: the Arrow tool, click on points A or B, and keeping the left button depressed, drag it. Observe the values that remain equal as you do so.

12. Select: the Text tool, double click in the blank region to open a dialog box and (1) explain what you learned about the angles and the sides of an equilateral triangle, (2) name the specific value to which all three angles of the triangle are equal, (3) give your own definition of an equilateral triangle.

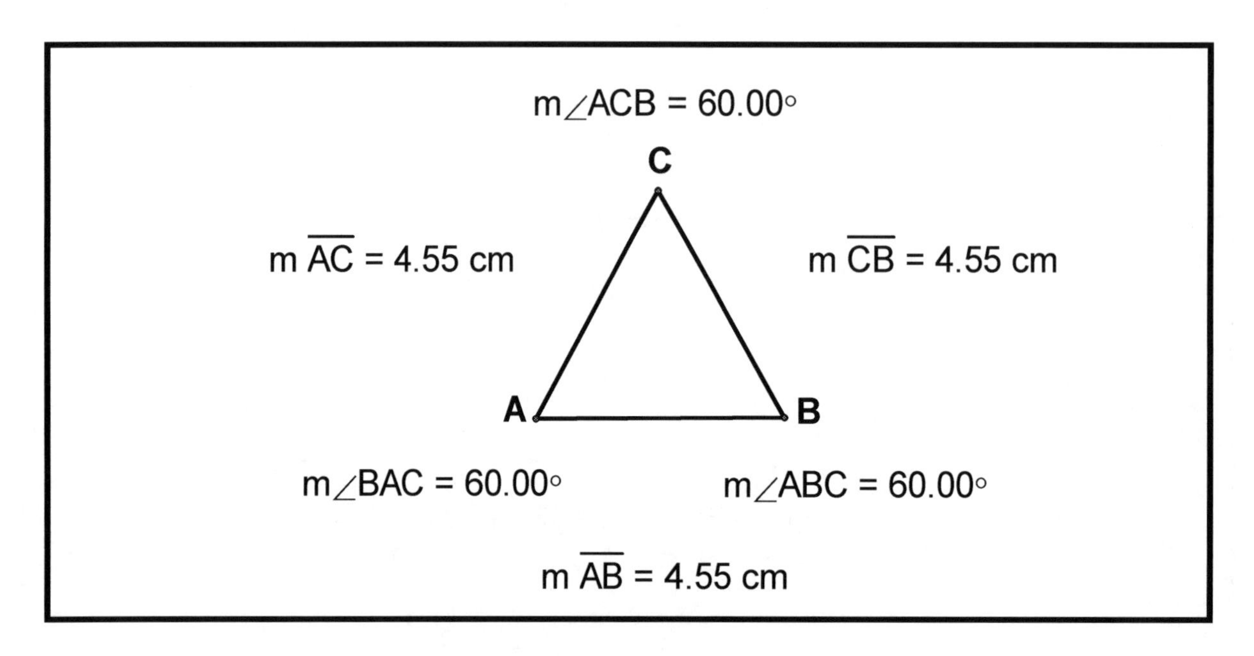

SUGGESTED EXERCISES

Type the solutions to the following problems in the blank region of your sketch.

1. ΔABC is an equilateral triangle. Find the measure of the sides $\overline{AC}$ and $\overline{BC}$ if the side $m\overline{AB} = 6$ cm.

2. ΔABC is an equilateral triangle. Find the value of x if $m\overline{AB} = 2x + 4$ and $m\overline{AC} = 10$.

3. ΔABC is an equilateral triangle. Find the value of x if $m\angle ABC = 2x - 10$.

4. ΔABC is an equilateral triangle. If $m\overline{AC} = 4x - 14$ and $m\overline{AB} = x + 22$, find the value of x and the measure of $\overline{AC}$.

LAB #23: THE TRIANGLE INEQUALITY

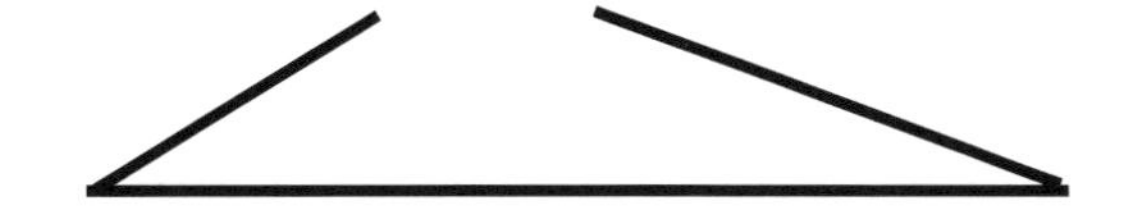

1. Select: the Point tool and plot three points as the vertices of a triangle.
2. Select: the Arrow tool, [EDIT], [SELECT ALL], [CONSTRUCT], [SEGMENTS] and deselect.
3. Select: the Text tool, bring the pointer to the middle of the lower line segment, see the hand turn black and click.
4. Label the two remaining line segments the same way.
5. Select: the Arrow tool, line segment j, [EDIT], [PROPERTIES], [LABEL], type a lower case m in the new window, click [OK] and deselect.
6. Do the same to rename the two remaining line segments as a lower case n and a lower case p.
7. Select: line segments m, n, p, [MEASURE], [LENGTHS], and deselect.
8. Select: [NUMBER], [CALCULATE], click on the caption that shows the measure of line segment m, [+], click on the caption that shows the measure of line segment n, and [OK].
9. Click on the caption that shows the value of (m + n), drag it to the right of the caption that shows the measure of line segment p, and deselect.
10. Select: [NUMBER], [CALCULATE], click on the caption that shows the measure of line segment m, [+], click on the caption that shows the measure of line segment p, and [OK].
11. Click on the caption that shows the value of (m + p), drag it to the right of the caption that shows the measure of line segment n, and deselect.
12. Select: [NUMBER], [CALCULATE], click on the caption that shows the measure of line segment n, [+], click on the caption that shows the measure of line segment p, and [OK].
13. Click on the caption that shows the value of (n + p), drag it to the right of the caption that shows the measure of the line segment m, and deselect.
14. Select: the Text tool, double click in the blank region to open a dialog box and describe your observations about the value of the sum of any two sides of a triangle and the measure of the remaining side.

15. Select: the Arrow tool, click on the vertex of the triangle between sides m and n, and keeping the left button depressed, drag it towards side p until the triangle degenerates into a single line segment. Observe the value of the sum of the two sides m and n as you get the vertex near side p.
16. Select: the Text tool, double click in the blank region to open a dialog box and explain what happens to the triangle if the sum of two sides becomes equal to the measure of the third side.

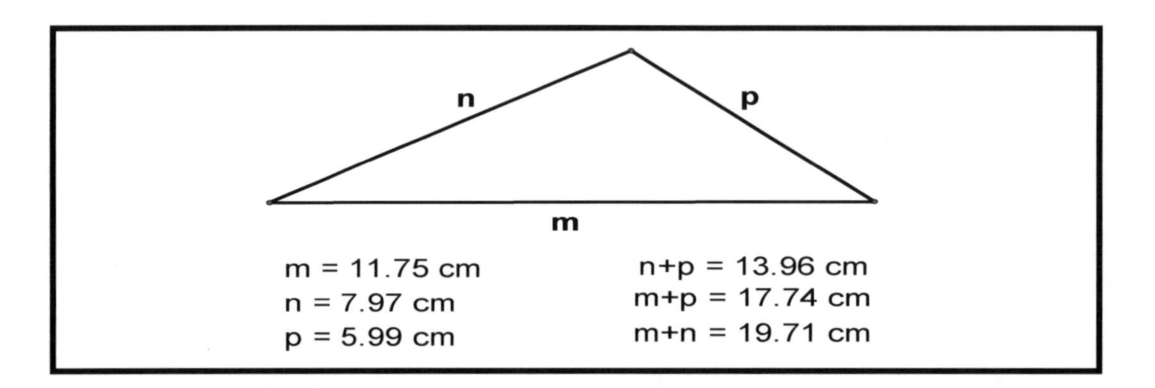

SUGGESTED EXERCISES

Type the solutions to the following problems in the blank region of your sketch.

1. Is it possible for a triangle to have sides with the given lengths: 3 ft, 6 ft, and 10 ft? Explain your answer.

2. Which set of numbers can represent the lengths of the sides of a triangle: (1) {2, 4, 6} (2) {2, 4, 4} (3) {4, 4, 8} (4) {4, 6, 12}. Explain your answer.

3. If two given sides of a triangle measure respectively 2 in and 4 in, find <u>all</u> possible integer lengths for the third side. Explain.

LAB #24: THE PYTHAGOREAN THEOREM

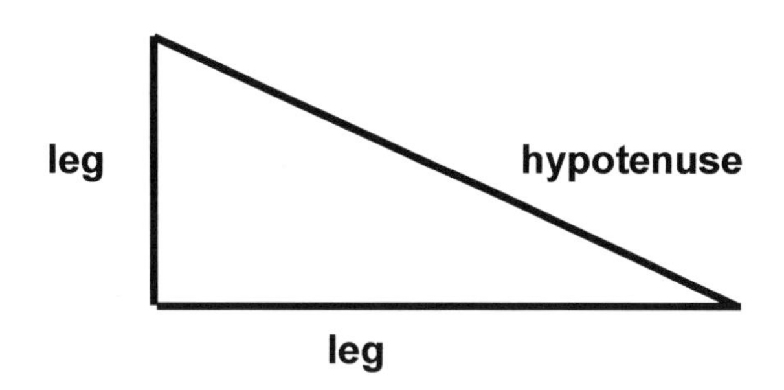

1. Select: [EDIT], [PREFERENCES], [TEXT], check the box [FOR ALL NEW POINTS] and click [OK].
2. Select: the Line Segment tool and construct a horizontal line segment.
3. Select: the Arrow tool and click in the blank region to deselect.
4. Select: point B, line segment $\overline{AB}$ [CONSTRUCT], [PERPENDICULAR LINE], [CONSTRUCT] and [POINT ON PERPENDICULAR LINE].
5. Click on point C and keeping the left button depressed, drag it one inch above point B and deselect.
6. Select: line $\overleftrightarrow{BC}$, [DISPLAY], and [HIDE PERPENDICULAR LINE].
7. Select: points A, then B, then C, [CONSTRUCT], [SEGMENTS], [MEASURE], [LENGTHS] and deselect.
8. Select: points A, then B, then C, [MEASURE], [ANGLE] and deselect.
9. Select: [NUMBER], [CALCULATE], click on the caption that shows $m\overline{CA}$, [∧], [2], [OK] and deselect.
10. Select: [NUMBER], [CALCULATE], click on the caption that shows $m\overline{AB}$, [∧], [2], [+], click on the caption that shows $m\overline{BC}$, [∧], [2], [OK] and deselect.

11. Click on any vertex of the triangle, and keeping the left button depressed, drag the triangle into different sizes. Observe which values remain equal as you do so.

12. Select: the Text tool, double click in the blank region to open a dialog box, and (1) name the side that represents the hypotenuse, (2) name the sides that represent the legs of the right triangle, and (3) explain the relationship between the sum of the squares of the legs of the right triangle and the square of the hypotenuse.

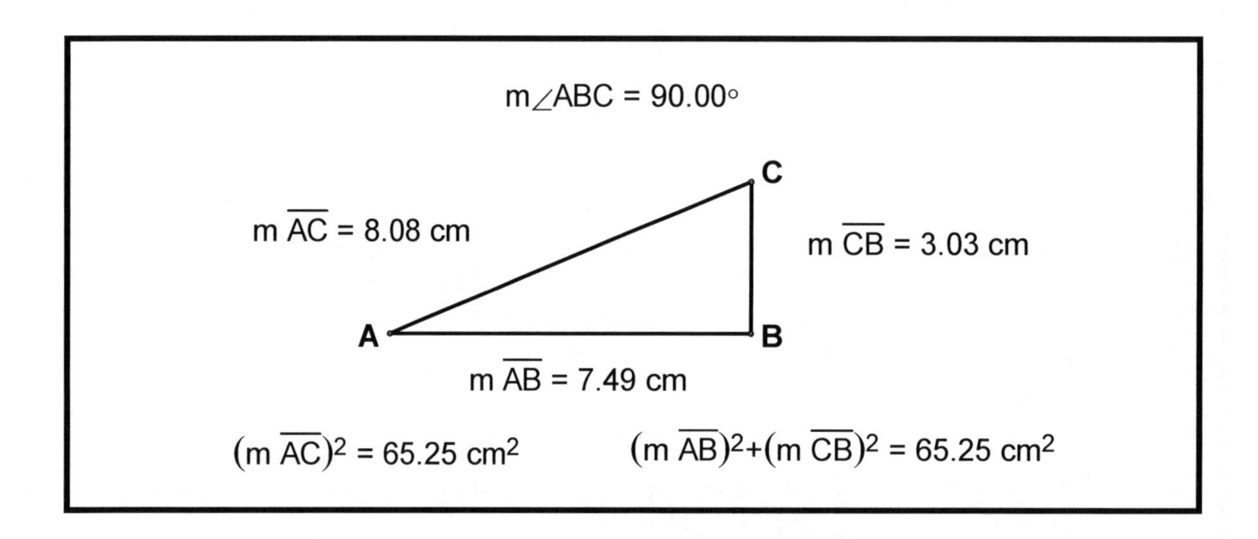

SUGGESTED EXERCISES

Type the solutions to the following problems in the blank region of your sketch.

1. The legs of a right triangle are 3 cm and 4 cm. Find the measure of the hypotenuse of the triangle.

2. Find the measure of longer leg of the right triangle if the measure of its hypotenuse is 13 in and the measure of the shorter leg is 5 in.

3. ΔABC is a right triangle. If the measure of leg $\overline{AB}$ is equal to x, and the measure of leg $\overline{BC}$ is equal to 2x, and the measure of hypotenuse $\overline{AC}$ = 10, find the value of x, correct to the nearest tenth.

LAB #25: THE PROPERTIES OF REGULAR POLYGONS

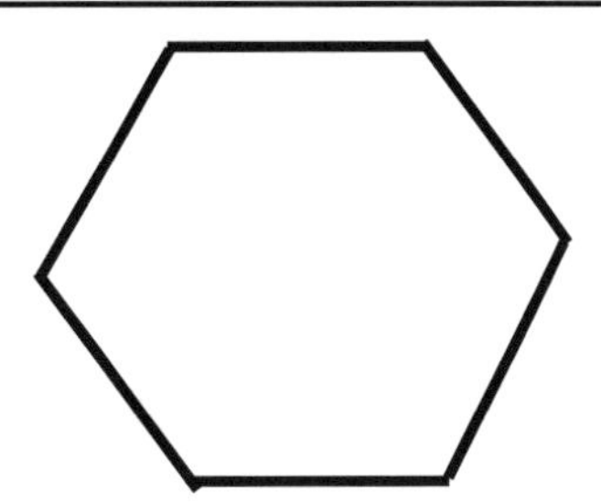

1. Select: [EDIT], [PREFERENCES], [TEXT], check the box [FOR ALL NEW POINTS] and click [OK].
2. Select: the Point tool and plot two points approximately one inch apart from each other in the middle of your screen.
3. Select: the Arrow tool and click in the blank region to deselect.
4. Select: point B, [TRANSFORM] and [MARK CENTER].
5. Select: point A, [TRANSFORM] and [ROTATE].
6. Type 60° in the display region named [DEGREES] and click [ROTATE]. NOTE: ***KEEP THE NEW POINT SELECTED!***
7. Select: [TRANSFORM], [ROTATE] and [ROTATE].
8. Select: [TRANSFORM], [ROTATE] and [ROTATE].
9. Select: [TRANSFORM], [ROTATE] and [ROTATE].
10. Select: [TRANSFORM], [ROTATE], [ROTATE] and deselect.
11. Select: point B, [DISPLAY] and [HIDE POINT].
12. Select: point A', [EDIT], [PROPERTIES], [LABEL], type a capital B, in the displayed window, click [OK] and deselect.
13. Select: point B', [EDIT], [PROPERTIES], type a capital C, in the displayed window, click [OK] and deselect.
14. Select: point C', [EDIT], [PROPERTIES], type a capital D, in the displayed window, click [OK] and deselect.
15. Select: point D', [EDIT], [PROPERTIES], type a capital E, in the displayed window, click [OK] and deselect.
16. Select: point E', [EDIT], [PROPERTIES], type a capital F, in the displayed window and click [OK].
17. Select: [EDIT], [SELECT ALL], [CONSTRUCT], [SEGMENTS], [MEASURE], [LENGTHS] and deselect.

18. Select: the Text tool, double click in the blank region to open a dialog box and describe your observations about the measures of the sides of a regular hexagon.
19. Select: the Arrow tool and click in the blank region to deselect.
20. Select: points A, then B, then C, [MEASURE], [ANGLE] and deselect.
21. Repeat the same to measure $\angle BCD$, $\angle CDE$, $\angle DEF$, $\angle EFA$, and $\angle FAB$.
22. Select: the Text tool, double click in the blank region to open a dialog box and describe your observations about the measures of the angles of the constructed hexagon.
23. Select: the Arrow tool, click on point B or C, and keeping the left button depressed, drag it. Observe what remains equal as you do so.
24. Select the Text tool, double click in the blank region to open a dialog box, and explain why this constructed hexagon is called regular. Give your own definition of a regular polygon having any number of given sides.

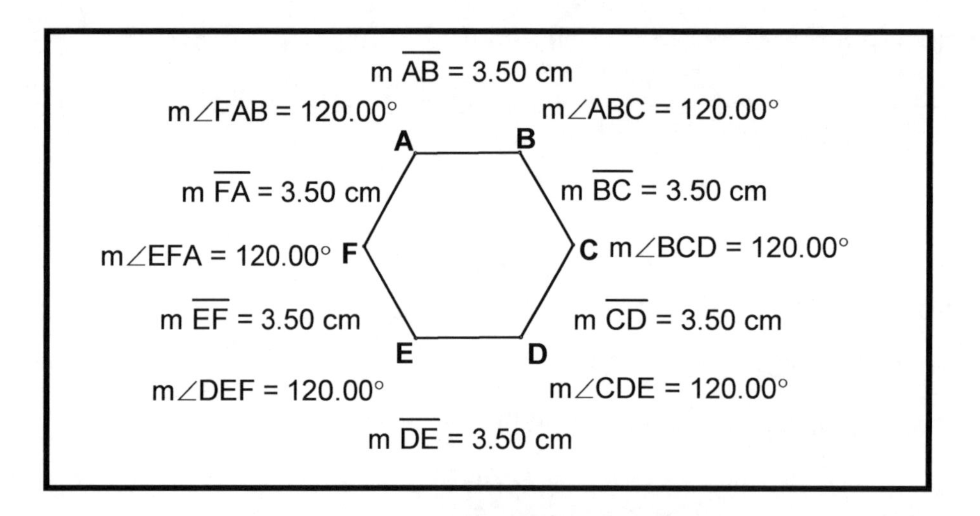

SUGGESTED EXERCISES

Type the solutions to the following problems in the blank region of your sketch. Choose the correct answer(s) in each of the following:

1. A regular polygon can have: (1) all congruent angles and all congruent sides (2) all congruent angles and no congruent sides (3) no congruent angles and no congruent sides.

2. Which of the following choices are regular polygons? (1) isosceles triangle, (2) rectangle, (3) square, (4) rhombus, (5) equilateral triangle.

3. One side of a regular hexagon is 13.2. Find each of the other sides.

LAB #26: THE OPPOSITE ANGLES OF A PARALLELOGRAM

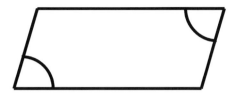

1. Select: [EDIT], [PREFERENCES], [TEXT], check the box [FOR ALL NEW POINTS] and click [OK].
2. Select: the Line Segment tool and construct a horizontal line segment in the middle of your screen.
3. Select: the Point tool and plot one point anywhere above line segment $\overline{AB}$.
4. Select: the Arrow tool and click in the blank region to deselect.
5. Select: points B, C, [CONSTRUCT], [SEGMENT] and deselect.
6. Select: point A, line segment $\overline{BC}$, [CONSTRUCT], [PARALLEL LINE] and deselect.
7. Select: point C, line segment $\overline{AB}$, [CONSTRUCT], [PARALLEL LINE] and deselect.
8. Select: the parallel line passing through A, the parallel line passing through C, [CONSTRUCT], [INTERSECTION] and deselect.
9. Select: line $\overleftrightarrow{CD}$, line $\overleftrightarrow{AD}$, [DISPLAY] and [HIDE PARALLEL LINES].
10. Select: points A, then D, then C, then B, [CONSTRUCT], [SEGMENTS] and deselect.
11. Select: points B, then A, then D, [MEASURE], [ANGLE] and deselect.
12. Do the same to measure $\angle BCD$.
13. Select: the Text tool, double click in the blank region to open a dialog box and describe your observations about the measures of the opposite angles $\angle BAD$ and $\angle BCD$.
14. Select: the Arrow tool and click in the blank region to deselect.
15. Select: points A, then D, then C, [MEASURE], [ANGLE] and deselect.

16. Do the same to measure ∠ABC.
17. Select: the Text tool, double click in the blank region to open a dialog box and describe your observations about the measures of the opposite angles ∠ADC and ∠ABC.
18. Select: the Arrow tool, click on vertex A, B, or C, and keeping the left button depressed, drag it. Observe what angle values remain the same as you do so.
19. Select: the Text tool, double click in the blank region to open the dialog box and explain what you have learned about the opposite angles of a parallelogram.

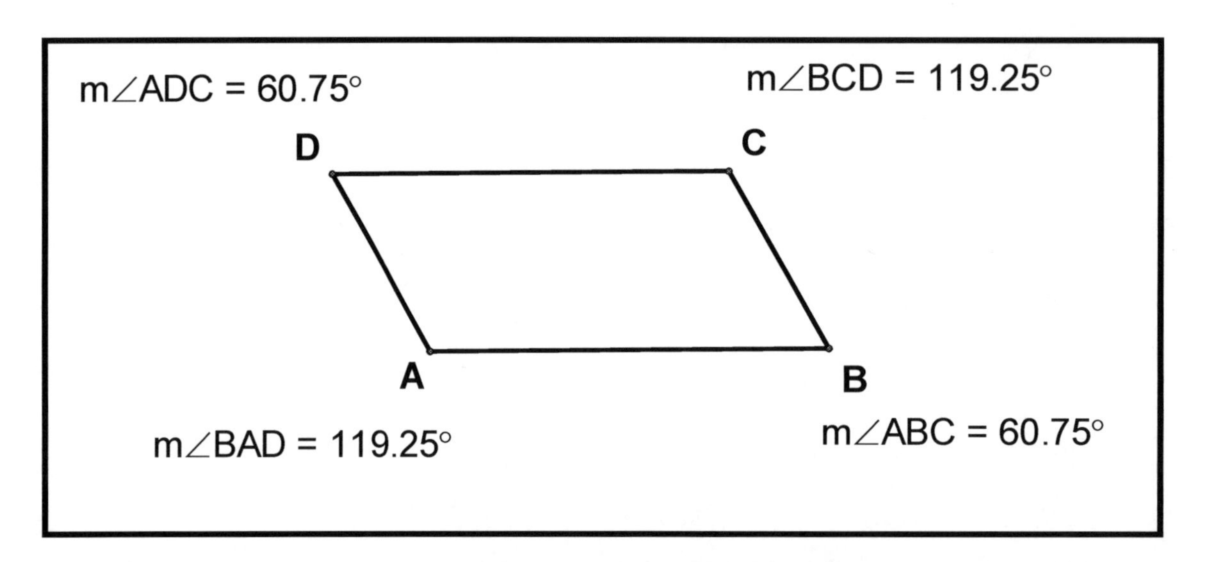

<u>SUGGESTED EXERCISES</u>

Type the solutions to the following problems in the blank region of your sketch.

1. ABCD is a parallelogram. If m∠A = 49°, find the measure of the opposite ∠C.

2. ABCD is a parallelogram. If m∠D = 3x and m∠B = 123°, find the value of x.

3. ABCD is a parallelogram. If m∠ ABC = 2x + 50 and m∠ADC = 3x + 40. Find the value of x and the measure of each angle.

LAB #27: THE CONSECUTIVE ANGLES OF A PARALLELOGRAM

1. Select: [EDIT], [PREFERENCES], [TEXT], check the box [FOR ALL NEW POINTS] and click [OK].
2. Select: the Line Segment tool and construct a horizontal line segment in the middle of your screen.
3. Select: the Point tool and plot one point anywhere above line segment $\overline{AB}$.
4. Select: the Arrow tool and click in the blank region to deselect.
5. Select: points B, C, [CONSTRUCT], [SEGMENT] and deselect.
6. Select: point A, line segment $\overline{BC}$, [CONSTRUCT], [PARALLEL LINE] and deselect.
7. Select: point C, line segment $\overline{AB}$, [CONSTRUCT], [PARALLEL LINE] and deselect.
8. Select: the parallel line passing through A, the parallel line passing through C, [CONSTRUCT], [INTERSECTION] and deselect.
9. Select: line $\overleftrightarrow{CD}$, line $\overleftrightarrow{AD}$, [DISPLAY] and [HIDE PARALLEL LINES].
10. Select: points A, then D, then C, then B, [CONSTRUCT], [SEGMENTS] and deselect.
11. Select: points A, then B, then C, [MEASURE], [ANGLE] and deselect.
12. Do the same to measure $\angle$BAD.
13. Select: [NUMBER], [CALCULATE], click on the caption that shows m$\angle$ABC, [+], click on the caption that shows m$\angle$BAD and [OK].
14. Click on point A, B, or C, and keeping the left button depressed, drag it. Observe the value that stays the same as you do so.
15. Select: the Text tool, double click in the blank region to open

a dialog box and describe your observations about the sum of these two consecutive angles.

16. Select: the Arrow tool and click in the blank region to deselect.
17. Select: points A, then D, then C, [MEASURE], [ANGLE] and deselect.
18. Do the same to measure ∠BCD.
19. Select: [NUMBER], [CALCULATE], click on the caption that shows m∠ADC, [+], click on the caption that shows m∠BCD, [OK] and deselect.
20. Find the sum of two more pairs of the consecutive angles in the parallelogram.
21. Click on point A, B, or C, and keeping the left button depressed, drag it. Observe what remains the same as you do so.
22. Select: the Text tool, double click in the blank region to open a dialog box and explain what you have learned about the sum of any pair of consecutive angles of a parallelogram.

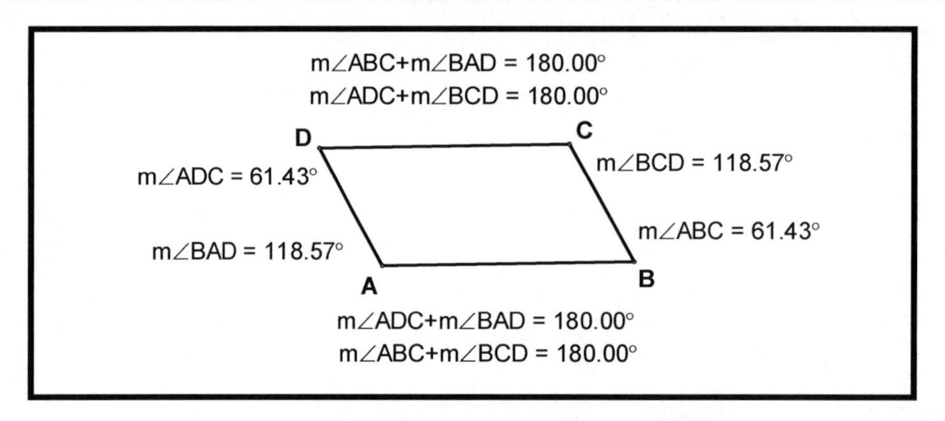

SUGGESTED EXERCISES

Type the solutions to the following problems in the blank region of your sketch.

1. ∠A and ∠B are consecutive angles of a parallelogram ABCD. If m∠A = 37°, find the m∠B.

2. ABCD is a parallelogram. If m∠BAD = 50° and m∠ABC = 4x, find the value of x.

3. ABCD is a parallelogram. If m∠BAD = 2x + 10 and m∠ABC = 3x + 20, find the value of x and the measure of each of the four angles of the parallelogram.

LAB #28: THE OPPOSITE SIDES OF A PARALLELOGRAM

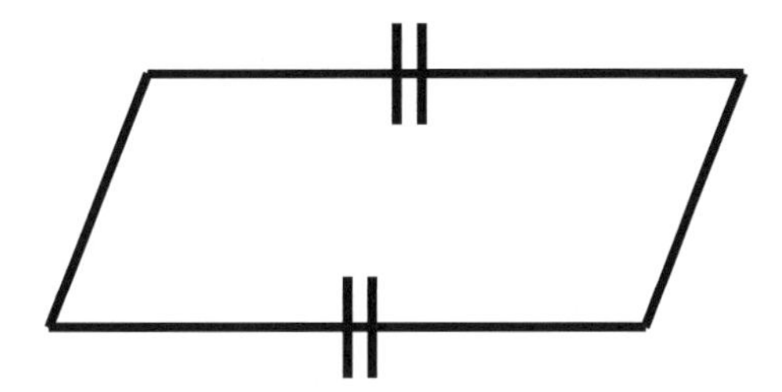

1. Select: [EDIT], [PREFERENCES], [TEXT], check the box [FOR ALL NEW POINTS] and click [OK].
2. Select: the Line Segment tool and construct a horizontal line segment in the middle of your screen.
3. Select: the Point tool and plot one point anywhere above line segment $\overline{AB}$.
4. Select: the Arrow tool and click in the blank region to deselect.
5. Select: points B, C, [CONSTRUCT], [SEGMENT] and deselect.
6. Select: point A, line segment $\overline{BC}$, [CONSTRUCT], [PARALLEL LINE] and deselect.
7. Select: point C, line segment $\overline{AB}$, [CONSTRUCT], [PARALLEL LINE] and deselect.
8. Select: the parallel line passing through A, the parallel line passing through C, [CONSTRUCT], [INTERSECTION] and deselect.
9. Select: line $\overleftrightarrow{CD}$, line $\overleftrightarrow{AD}$, [DISPLAY] and [HIDE PARALLEL LINES].
10. Select: points A, then D, then C, then B, [CONSTRUCT], [SEGMENTS] and deselect.
11. Select: side $\overline{AB}$, side $\overline{CD}$, [MEASURE] and [LENGTHS].
12. Select: the Text tool, double click in the blank region to open a dialog box and describe your observations about the measures of these two opposite sides.
13. Select: the Arrow tool and click in the blank region to deselect.
14. Select: sides $\overline{BC}$, $\overline{AD}$, [MEASURE], [LENGTHS] and deselect.

15. Select: the Text tool, double click in the blank region to open a dialog box and describe your observations about the measures of these two opposite sides.
16. Click on point B, and keeping the left button depressed, drag it. Observe which line segments remain congruent as you do so.
17. Select: the Text tool, double click in the blank region to open a dialog box and explain what you learned about the opposite sides of a parallelogram.

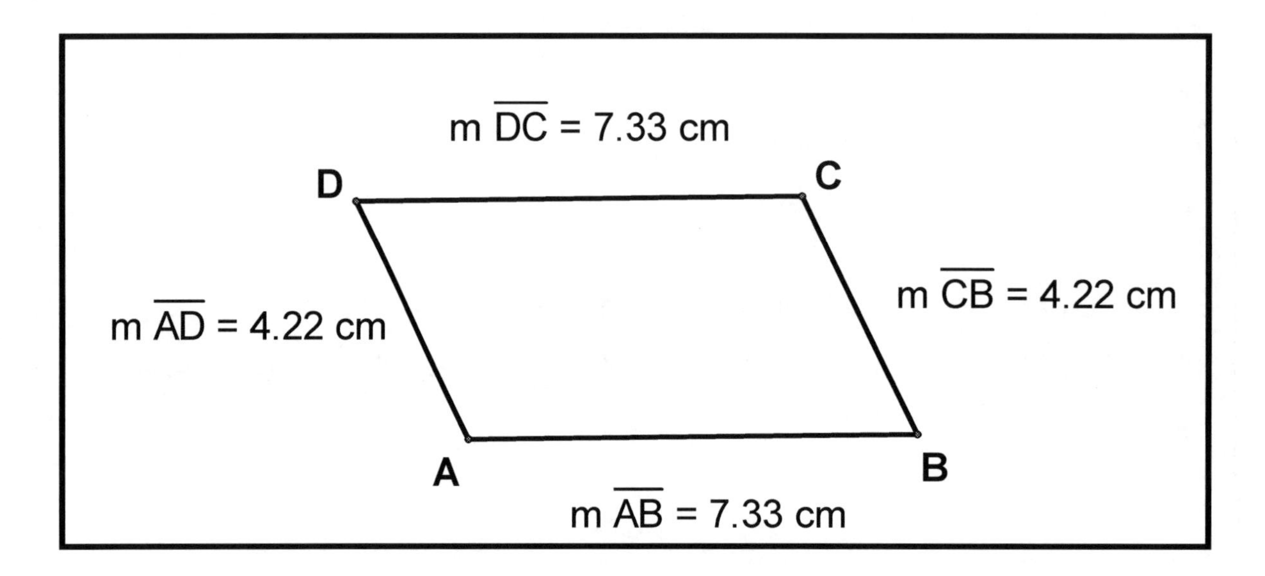

SUGGESTED EXERCISES

Type the solutions to the following problems in the blank region of your sketch.

1. ABCD is a parallelogram. If the measure of side $\overline{AB}$ is 12 cm, find the measure of side $\overline{CD}$.

2. ABCD is a parallelogram. If the measure of side $\overline{AD}$ = 62 cm, and the measure of side $\overline{BC}$ = 5x, find the value of x.

3. ABCD is a parallelogram. If the lengths of side $\overline{AB}$ is represented as 3x + 8 and $\overline{CD}$ as x + 42, find the value of x.

LAB # 29: THE DIAGONALS OF A PARALLELOGRAM

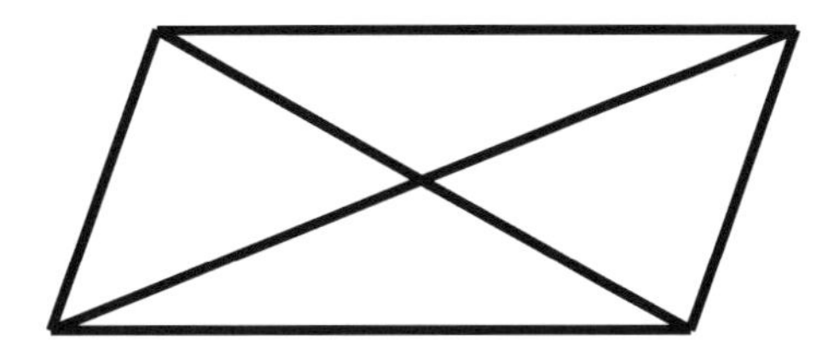

1. Select: [EDIT], [PREFERENCES], [TEXT], check the box [FOR ALL NEW POINTS] and click [OK].
2. Select: the Line Segment tool and construct a horizontal line segment in the middle of your screen.
3. Select: the Point tool and plot one point anywhere above line segment $\overline{AB}$.
4. Select: the Arrow tool and click in the blank region to deselect.
5. Select: points B, C, [CONSTRUCT], [SEGMENT] and deselect.
6. Select: point A, line segment $\overline{BC}$, [CONSTRUCT], [PARALLEL LINE] and deselect.
7. Select: point C, line segment $\overline{AB}$, [CONSTRUCT], [PARALLEL LINE] and deselect.
8. Select: the parallel line passing through A, the parallel line passing through C, [CONSTRUCT], [INTERSECTION] and deselect.
9. Select: line $\overleftrightarrow{CD}$, line $\overleftrightarrow{AD}$, [DISPLAY] and [HIDE PARALLEL LINES].
10. Select: points A, then D, then C, then B, [CONSTRUCT], [SEGMENTS] and deselect.
11. Select: points A, then C, then B, then D, [CONSTRUCT], [SEGMENTS] and deselect.
12. Select: the Arrow tool, diagonals $\overline{AC}$ and $\overline{BD}$, [CONSTRUCT], [INTERSECTION] and deselect.
13. Select: points A, then E, then C, [CONSTRUCT], [SEGMENTS], [MEASURE], [LENGTHS] and deselect.
14. Select: points B, then E, then D, [CONSTRUCT], [SEGMENTS], [MEASURE], [LENGTHS] and deselect.

15. Select: the Text tool, double click in the blank region to open a dialog box and (1) describe your observations about the measures of the diagonals $\overline{AC}$ and $\overline{BD}$, (2) describe your observations about the line segments $\overline{AE}$ and $\overline{EC}$, $\overline{BE}$ and $\overline{ED}$, (3) explain what the diagonals of the parallelogram do to each other.
16. Select: the Arrow tool and click in the blank region to deselect.
17. Select: points A, then C, then D, [MEASURE], [ANGLE] and deselect.
18. Do the same to measure $\angle ACB$.
19. Select: the Text tool, double click in the blank region to open a dialog box and write if the diagonals of the parallelogram bisect (divide in half) the angles of the parallelogram.
20. Select: the Arrow tool, click on point C, and keeping the left button depressed, drag it. Observe the measures that remain equal as you do so.
21. Select: the Text tool, double click in the blank region to open a dialog box and explain what you have learned about the diagonals of the parallelogram.

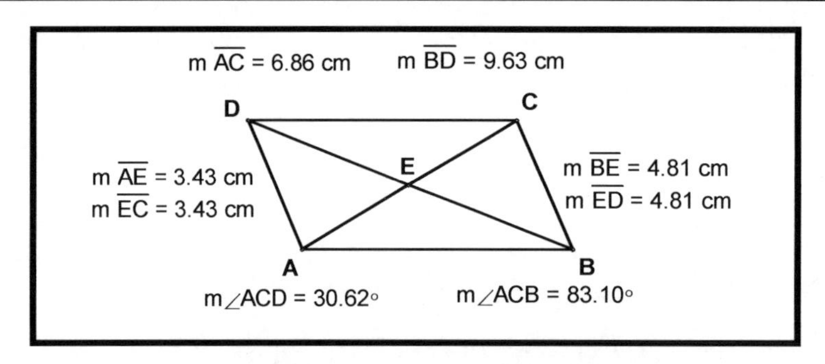

SUGGESTED EXERCISES

Type the solutions to the following problems in the blank region of your sketch.

1. $\overline{AC}$ and $\overline{BD}$ are the diagonals of parallelogram ABCD. The point of intersection of two diagonals is E. If line segment $\overline{AE}$ is equal to 4 cm, find the measure of line segment $\overline{CE}$.

2. $\overline{AC}$ and $\overline{BD}$ are the diagonals of parallelogram ABCD. The point of intersection of two diagonals is E. If $m\overline{CE} = 6x$ and $m\overline{AE} = 42$ cm, find the value of x.

3. $\overline{AC}$ and $\overline{BD}$ are the diagonals of parallelogram ABCD. If $m\overline{BE} = 5x$ and $m\overline{DE} = 3x + 7$, find the value of x and the measure of each line segment.

LAB #30: THE PROPERTIES OF A RECTANGLE

1. Select: [EDIT], [PREFERENCES], [TEXT], check the box [FOR ALL NEW POINTS] and click [OK].
2. Select: the Line Segment tool and construct a horizontal line segment in the middle of your screen.
3. Select: the Arrow tool, [EDIT], [SELECT ALL], [CONSTRUCT], [PERPENDICULAR LINES] and deselect.
4. Select: the perpendicular line passing through B, [CONSTRUCT] and [POINT ON PERPENDICULAR LINE].
5. Click on point C, and keeping the left button depressed, drag it approximately one inch above point B and deselect.
6. Select: point C, line $\overleftrightarrow{BC}$, [CONSTRUCT], [PERPENDICULAR LINE] and deselect.
7. Select: the perpendicular line passing through A, the perpendicular line passing through C, [CONSTRUCT], [INTERSECTION] and deselect.
8. Select: lines $\overleftrightarrow{AD}$, $\overleftrightarrow{CD}$, $\overleftrightarrow{BC}$, [DISPLAY] and [HIDE PERPENDICULAR LINES].
9. Select: points A, then D, then C, then B, [CONSTRUCT], [SEGMENTS] and deselect.
10. Select: points A, then B, then C, [MEASURE], [ANGLE] and deselect.
11. Do the same to measure $\angle BCD$, $\angle ADC$, and $\angle BAD$.
12. Select: the Text tool, double click in the blank region to open a dialog box and write your observations about the measures of the four angles of a rectangle.
13. Select: the Arrow tool and click in the blank region to deselect.
14. Select: sides $\overline{AB}$, $\overline{CD}$, [MEASURE] and [LENGTHS].

15. Select: the Text tool, double click in the blank region to open a dialog box and describe your observations about the measures of these two opposite sides.
16. Select: the Arrow tool and click in the blank region to deselect.
17. Select: sides $\overline{AD}$, $\overline{BC}$, [MEASURE] and [LENGTHS].
18. Select: the Text tool, double click in the blank region to open a dialog box and describe your observations about the measures of these two opposite sides.
19. Select: the Arrow tool, click on point A or B, and keeping the left button depressed, drag it. Observe what remains equal as you do so.
20. Select: the Text tool, double click in the blank region to open a dialog box and give your own definition of the rectangle.

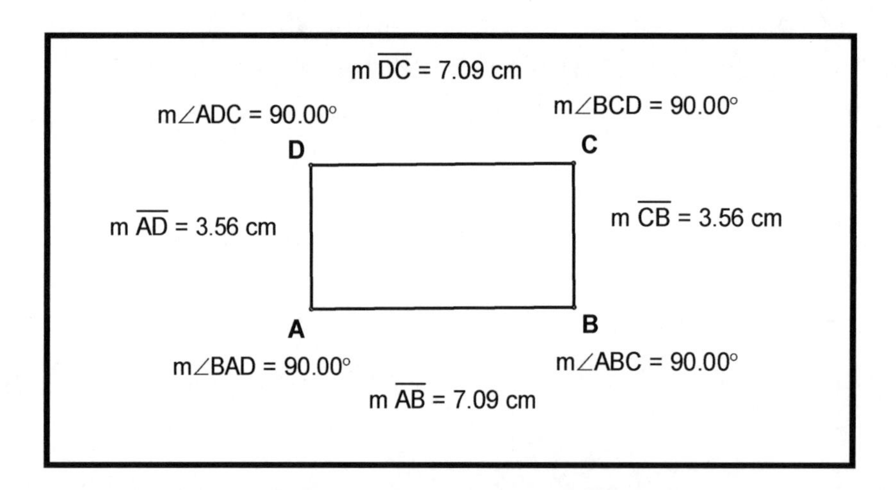

<u>SUGGESTED EXERCISES</u>

Type the solutions to the following problems in the blank region of your sketch.

1. ABCD is a rectangle. If $m\overline{AB} = 7$ cm, find the measure of side $\overline{CD}$.

2. ABCD is a rectangle. If $m\overline{AD} = 4x$ and $m\overline{BC} = 36$ cm, find the value of x.

3. ABCD is a rectangle. If $m\angle ADC = 5x - 10$, find x.

LAB #31: THE DIAGONALS OF A RECTANGLE

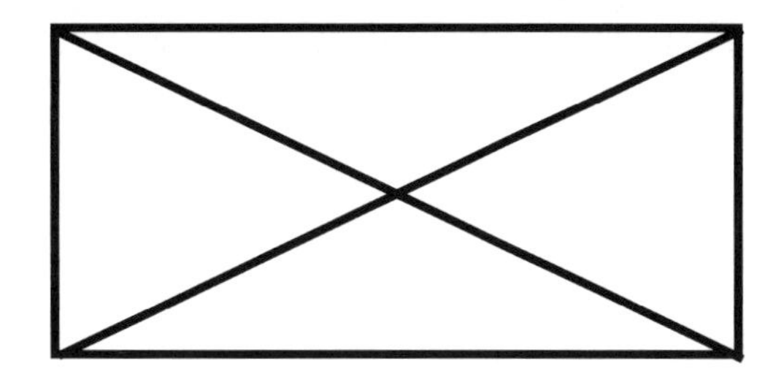

1. Select: [EDIT], [PREFERENCES], [TEXT], check the box [FOR ALL NEW POINTS] and click [OK].
2. Select: the Line Segment tool and construct a horizontal line segment in the middle of your screen.
3. Select: the Arrow tool, [EDIT], [SELECT ALL], [CONSTRUCT], [PERPENDICULAR LINES] and deselect.
4. Select: the perpendicular line passing through point B, [CONSTRUCT] and [POINT ON PERPENDICULAR LINE].
5. Click on point C, and keeping the left button depressed, drag it approximately one inch above point B and deselect.
6. Select: point C, line $\overleftrightarrow{BC}$, [CONSTRUCT], [PERPENDICULAR LINE] and deselect.
7. Select: the perpendicular line passing through point A, the perpendicular line passing through point C, [CONSTRUCT], [INTERSECTION] and deselect.
8. Select: lines $\overleftrightarrow{AD}$, $\overleftrightarrow{CD}$, $\overleftrightarrow{BC}$, [DISPLAY] and [HIDE PERPENDICULAR LINES].
9. Select: points A, then D, then C, then B, [CONSTRUCT], [SEGMENTS] and deselect.
10. Select: points A, then C, then B, then D, [CONSTRUCT], [SEGMENTS] and deselect.
11. Select: diagonals $\overline{AC}$, $\overline{BD}$, [MEASURE] and [LENGTHS].
12. Select: the Text tool, double click in the blank region to open a dialog box and describe your observations about the measures of the diagonals $\overline{AC}$ and $\overline{BD}$ of a rectangle ABCD.
13. Select: the Arrow tool and click in the blank region to deselect.
14. Select: diagonals $\overline{AC}$ and $\overline{BD}$, [CONSTRUCT], [INTERSECTION] and deselect.

15. Select: points A, E, [MEASURE], [DISTANCE] and deselect.
16. Do the same to measure line segments $\overline{CE}$, $\overline{BE}$, and $\overline{DE}$.
17. Select: the Text tool, double click in the blank region to open a dialog box and (1) describe your observations about the line segments $\overline{AE}$ and $\overline{CE}$, $\overline{BE}$ and $\overline{DE}$, (2) explain what diagonals of the rectangle do to each other.
18. Select: the Arrow tool and click in the blank region to deselect.
19. Select: points A, C, D, [MEASURE], [ANGLE] and deselect.
20. Do the same to measure ∠ACB.
21. Select: the Text tool, double click in the blank region to open a dialog box and explain if diagonals of the rectangle bisect (divide in half) the angles of the rectangle.
22. Select: the Arrow tool, click on point C, and keeping the left button depressed, drag it. Observe the measures that remain equal as you do so.
23. Select: the Text tool, double click in the blank region to open a dialog box and explain what you have learned about the diagonals of the rectangle.

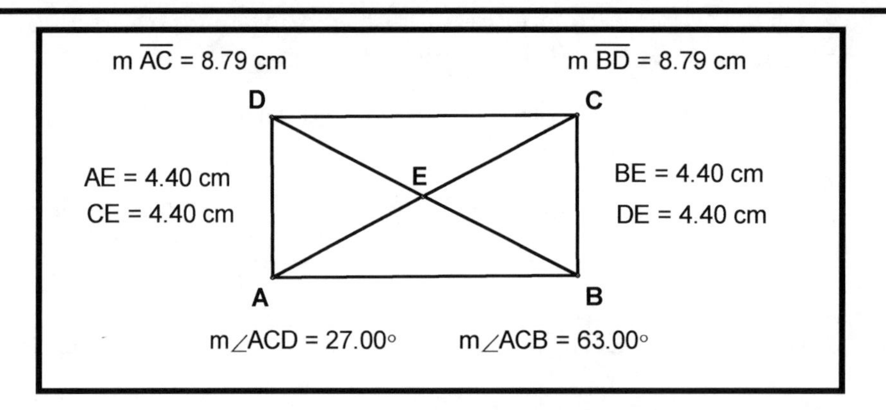

SUGGESTED EXERCISES

Type the solutions to the following problems in the blank region of your sketch.

1. ABCD is a rectangle. If m$\overline{AE}$ = 9 cm, find the measure of $\overline{CE}$.

2. ABCD is a rectangle. If m∠ACD = 4x + 6 and m∠ACB = 64°, find the value of x and the measure of ∠ACD.

3. $\overline{AC}$ and $\overline{BD}$ are the diagonals of rectangle ABCD. If m$\overline{AC}$ = 3x + 20 and m$\overline{BD}$ = x + 44, find x and the measure of each diagonal.

LAB #32: THE PROPERTIES OF A SQUARE

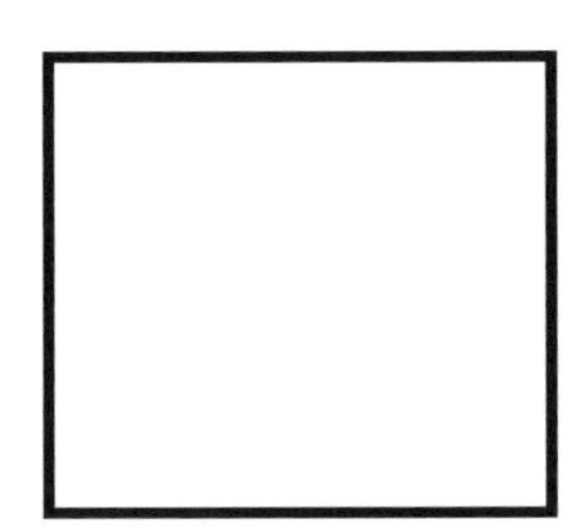

1. Select: [EDIT], [PREFERENCES], [TEXT], check the box [FOR ALL NEW POINTS] and click [OK].
2. Select: the Point tool and plot two points approximately one inch apart from each other in the middle of your screen.
3. Select: the Arrow tool and click in the blank region to deselect.
4. Select: point B, [TRANSFORM] and [MARK CENTER].
5. Select: point A, [TRANSFORM], [ROTATE], [ROTATE] and deselect.
6. Select: point A', [EDIT], [PROPERTIES], [LABEL], type a capital C in the displayed window, click [OK] and deselect.
7. Select: point C, [TRANSFORM] and [MARK CENTER].
8. Select: point B, [TRANSFORM], [ROTATE], [ROTATE] and deselect.
9. Select: point B', [EDIT], [PROPERTIES], type a capital D in the displayed window, click [OK] and deselect.
10. Select: [EDIT], [SELECT ALL], [CONSTRUCT], [SEGMENTS], [MEASURE], [LENGTHS] and deselect.
11. Select: the Text tool, double click in the blank region to open a dialog box and describe your observations about the measures of the sides of the square.
12. Select: the Arrow tool and click in the blank region to deselect.
13. Select: points A, then B, then C, [MEASURE], [ANGLE] and deselect.

14. Do the same to measure $\angle BCD$, $\angle ADC$ and $\angle BAD$.
15. Select: the Text tool, double click in the blank region to open a dialog box, and describe your observations about the measures of the angles of the square.
16. Select: the Arrow tool, click on point B, and keeping the left button depressed, drag it. Observe the measures of the angles and the sides of the square as you do so.
17. Select: the Text tool, double click in the blank region to open a dialog box and give your own definition of a square.

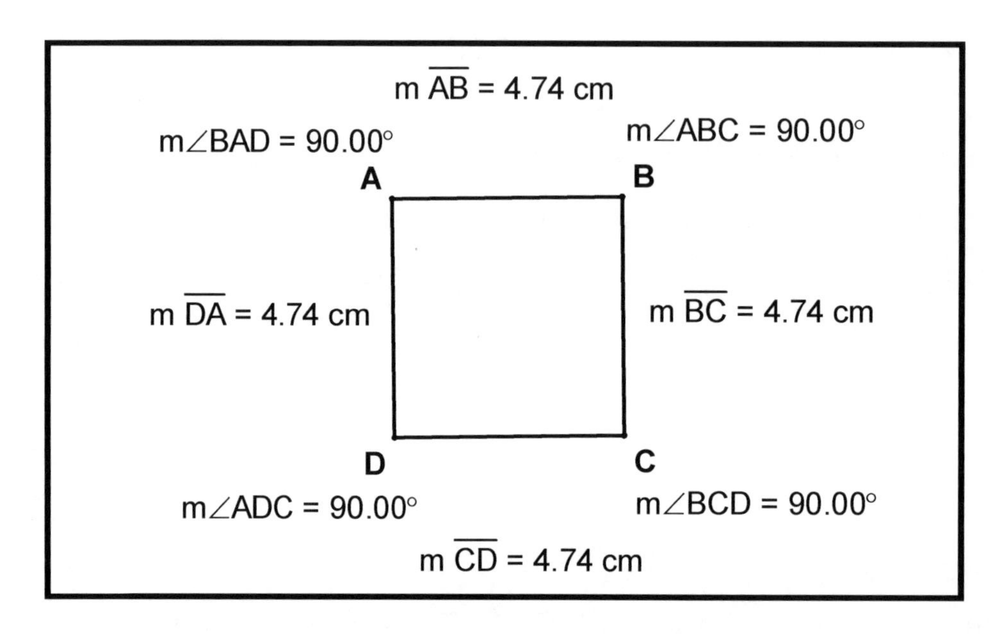

SUGGESTED EXERCISES

Type the solutions to the following problems in the blank region of your sketch.

1. ABCD is a square. If $m\overline{AB} = 5$ in, find the measure of $\overline{BC}$.

2. ABCD is a square. If $m\overline{AB} = 4x$ and $m\overline{BC} = 29$, find the value of x.

3. ABCD is a square. If $m\overline{AD} = 8x - 6$ and $m\overline{CD} = 5x + 12$, find the length of each side of the square.

4. ABCD is a square. If $m\angle A = 3x + 30$, find the value of x.

LAB #33: THE DIAGONALS OF A SQUARE

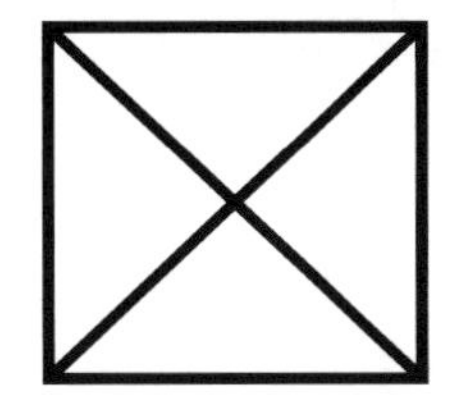

1. Select: [EDIT], [PREFERENCES], [TEXT], check the box [FOR ALL NEW POINTS] and click [OK].
2. Select: the Point tool and plot two points approximately one inch apart from each other in the middle of your screen.
3. Select: the Arrow tool and click in the blank region to deselect.
4. Select: point B, [TRANSFORM] and [MARK CENTER].
5. Select: point A, [TRANSFORM], [ROTATE], [ROTATE] and deselect.
6. Select: point A', [EDIT], [PROPERTIES], [LABEL], type a capital C in the displayed window, click [OK] and deselect.
7. Select: point C, [TRANSFORM] and [MARK CENTER].
8. Select: point B, [TRANSFORM], [ROTATE], [ROTATE] and deselect.
9. Select: point B', [EDIT], [PROPERTIES], type a capital D in the displayed window, click [OK] and deselect.
10. Select: [EDIT], [SELECT ALL], [CONSTRUCT], [SEGMENTS] and deselect.
11. Select: points A, then C, then B, then D, [CONSTRUCT], [SEGMENTS] and deselect.
12. Select: diagonals $\overline{AC}$, $\overline{BD}$, [MEASURE] and [LENGTHS].
13. Select: the Text tool, double click in the blank region to open a dialog box and describe your observations about the diagonals of the square.
14. Select: the Arrow tool and click in the blank region to deselect.
15. Select: diagonals $\overline{AC}$ and $\overline{BD}$, [CONSTRUCT], [INTERSECTION], [EDIT], [PROPERTIES], type a capital E in the displayed window, click [OK] and deselect.
16. Select: points A, E, [MEASURE], [DISTANCE] and deselect.
17. Do the same to measure line segments $\overline{CE}$, $\overline{BE}$, and $\overline{DE}$.
18. Select: the Text tool, double click in the blank region to open a dialog box and explain whether or not the diagonals of the square bisect each other (divide each other in half).

19. Select: the Arrow tool and click in the blank region to deselect.
20. Select: points A, then B, then D, [MEASURE], [ANGLE] and deselect.
21. Do the same to measure ∠CBD.
22. Select: the Text tool, double click in the blank region to open a dialog box and (1) describe your observations about the measures of ∠ABD and ∠CBD, (2) explain whether or the diagonals of the square bisect the angles of the square.
23. Select: the Arrow tool and click in the blank region to deselect.
24. Select: points A, then E, then B, [MEASURE], [ANGLE] and deselect.
25. Select: points C, then E, then D, [MEASURE], [ANGLE], and deselect.
26. Select: the Text tool, double click in the blank region to open a dialog box and explain if the diagonals of the square are perpendicular to each other. Justify your answer using the angle measures of ∠AEB and ∠CED.
27. Select: the Arrow tool, click on point B, and keeping the left button depressed, drag it. Observe the measures that remain equal as you do so.
28. Select: the Text tool, double click in the blank region to open a dialog box and explain what you have learned about the diagonals of the square.

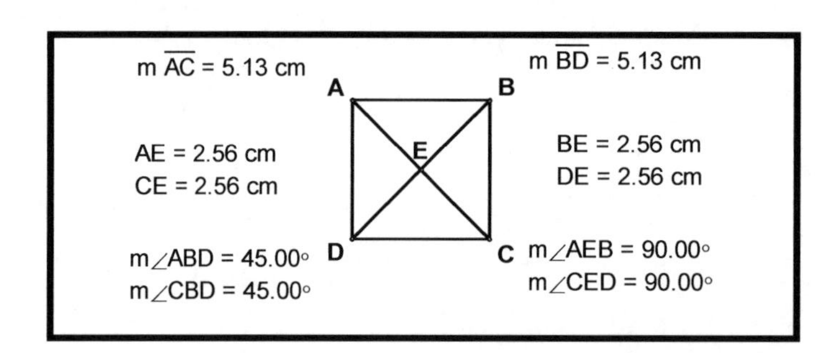

<u>SUGGESTED EXERCISES</u>

Type the solutions to the following problems in the blank region of your sketch.

1. $\overline{AC}$ and $\overline{BD}$ are the diagonals of square ABCD intersecting at E. If m$\overline{CE}$ = 5 cm, find the measure of $\overline{AE}$.

2. $\overline{AC}$ and $\overline{BD}$ are diagonals of the square ABCD intersecting at E. If m$\overline{AE}$ = 3x and m$\overline{CE}$ = 45 cm, find the value of x.

3. $\overline{AC}$ and $\overline{BD}$ are diagonals of the square ABCD intersecting at E. If m$\overline{BE}$ = 3x + 4, express the length of diagonal $\overline{BD}$ in terms of x.

4. In square ABCD, diagonal $\overline{BD}$ is drawn. If m∠ ADB = 2x, find x.

LAB #34: THE PROPERTIES OF A RHOMBUS

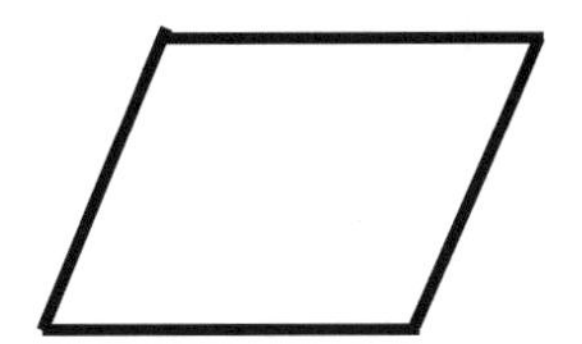

1. Select: [EDIT], [PREFERERNCES], [TEXT], check the box [FOR ALL NEW POINTS] and click [OK].
2. Select: the Point tool and plot two points approximately one inch apart from each other in the middle of your screen.
3. Select: the Arrow tool and click in the blank region to deselect.
4. Select: points B, A, [CONSTRUCT], [CIRCLE BY CENTER + POINT], [CONSTRUCT], [POINT ON CIRCLE] and deselect.
5. Select: points A, B, [CONSTRUCT], [SEGMENT] and deselect.
6. Select: points B, C, [CONSTRUCT], [SEGMENT] and deselect.
7. Select: point A, line segment $\overline{BC}$, [CONSTRUCT], [PARALLEL LINE] and deselect.
8. Select: line segment $\overline{AB}$, point C, [CONSTRUCT], [PARALLEL LINE] and deselect.
9. Select: the parallel line passing through A, the parallel line passing through C, [CONSTRUCT], [INTERSECTION] and deselect.
10. Select: the circle, lines $\overleftrightarrow{AD}$, $\overleftrightarrow{CD}$, [DISPLAY] and [HIDE PATH OBJECTS].
11. Select: points A, then B, then C, then D, [CONSTRUCT], [SEGMENTS], [MEASURE], [LENGTHS] and deselect.
12. Select: the Text tool, double click in the blank region to open the dialog box and describe your observations about the measures of the sides of the rhombus.
13. Select: the Arrow tool and click in the blank region to deselect.
14. Select: points A, then B, then C, [MEASURE], [ANGLE] and deselect.

15. Do the same to measure $\angle ADC$, $\angle BCD$, and $\angle BAD$.
16. Select: the Text tool, double click in the blank region to open a dialog box and describe your observations about the opposite angles of the rhombus ($\angle BAD$ and $\angle BCD$, $\angle ABC$ and $\angle ADC$.)
17. Select: the Arrow tool, click on point A or B, and keeping the left button depressed, drag it. Observe the values that remain equal as you do so.
18. Select: the Text tool, double click in the blank region to open a dialog box and explain how the rhombus is different from the square and give your own definition of the rhombus.

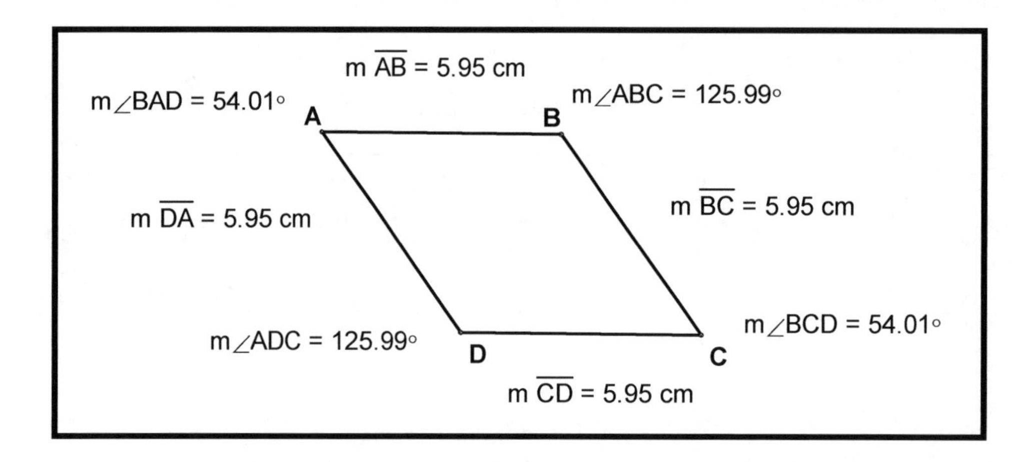

SUGGESTED EXERCISES

Type the solution to the following problems in the blank region of your sketch.

1. $\overline{AD}$ and $\overline{BC}$ are sides of rhombus ABCD. If $m\overline{AD} = 9$ cm, find the measure of the side $\overline{BC}$.

2. $\overline{AB}$ and $\overline{BC}$ are the sides of rhombus ABCD. If $m\overline{AB} = 3x$ and $m\overline{BC} = 48$ in, find the value of x.

3. $\overline{AD}$ and $\overline{BC}$ are the sides of the rhombus. If $m\overline{AD} = 3x - 15$ and $m\overline{BC} = x + 20$, find the value of x.

4. In rhombus ABCD, $m\angle BAD = 3x$ and $m\angle BCD = 2x + 50$. Find the value of x and $m\angle BAD$ and $m\angle BCD$.

LAB #35: THE DIAGONALS OF A RHOMBUS

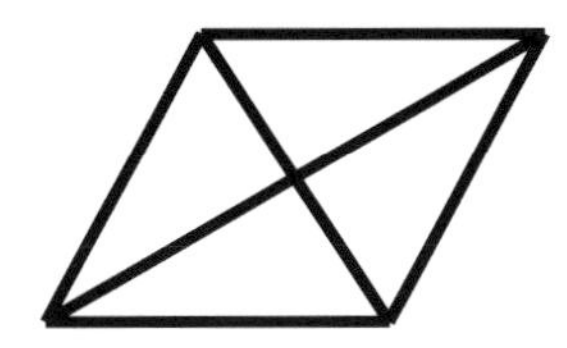

1. Select: [EDIT], [PREFERENCES], [TEXT], check the box [FOR ALL NEW POINTS] and click [OK].
2. Select: the Point tool and plot two points approximately one inch apart from each other in the middle of your screen.
3. Select: the Arrow tool and click in the blank region to deselect.
4. Select: points B, A, [CONSTRUCT], [CIRCLE BY CENTER + POINT], [CONSTRUCT], [POINT ON CIRCLE] and deselect.
5. Select: points A, B, [CONSTRUCT], [SEGMENT] and deselect.
6. Select: points B, C, [CONSTRUCT], [SEGMENT] and deselect.
7. Select: point A, line segment $\overline{BC}$, [CONSTRUCT], [PARALLEL LINE] and deselect
8. Select: line segment $\overline{AB}$, point C, [CONSTRUCT], [PARALLEL LINE] and deselect.
9. Select: the parallel line passing through A, the parallel line passing through C, [CONSTRUCT], [INTERSECTION] and deselect.
10. Select: the circle, lines $\overleftrightarrow{AD}$, $\overleftrightarrow{CD}$, [DISPLAY] and [HIDE PATH OBJECTS].
11. Select: points A, then D, then C, [CONSTRUCT], [SEGMENTS] and deselect.
12. Select: points B, D, [CONSTRUCT], [SEGMENT] and deselect.
13. Select: diagonals $\overline{AC}$, $\overline{BD}$, [MEASURE], [LENGTHS] and deselect.
14. Select: the Text tool, double click in the blank region to open a dialog box and write your observations about the measures of the diagonals of the rhombus.
15. Select: the Arrow tool and click in the blank region to deselect.
16. Select: diagonals $\overline{AC}$ and $\overline{BD}$, [CONSTRUCT], [INTERSECTION] and deselect.
17. Select: points A, E, [MEASURE], [DISTANCE] and deselect.
18. Do the same to measure line segments $\overline{CE}$, $\overline{BE}$, and $\overline{DE}$.
19. Select: the Text tool, double click in the blank region to open a dialog box and explain if the diagonals of the rhombus bisect each other.

20. Select: the Arrow tool and click in the blank region to deselect.
21. Select: points A, then B, then D, [MEASURE], [ANGLE] and deselect.
22. Do the same to measure ∠CBD.
23. Select: the Text tool, double click to open a dialog box and explain if diagonals bisect (divide in half) the angles of the rhombus.
24. Select: the Arrow tool and click in the blank region to deselect
25. Measure ∠AEB and ∠BEC.
26. Select: the Text tool, double click in the blank region to open a dialog box and explain if diagonals of the rhombus are perpendicular to each other. Justify your answer using the angle measure of ∠AEB and ∠BEC.
27. Select: the Arrow tool, click on point B, and keeping the left button depressed, drag it. Observe the measures that remain equal as you do so.
28. Select: the Text tool, double click in the blank region to open a dialog box and explain what you have learned about the diagonals of the rhombus.

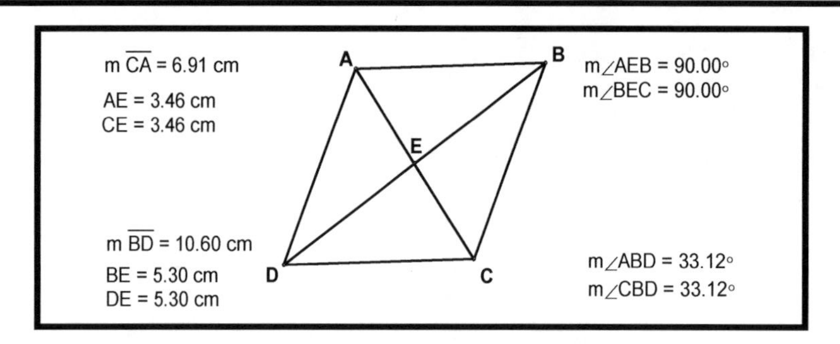

SUGGESTED EXERCISES

Type the solutions to the following problems in the blank region of your sketch.

1. $\overline{AC}$ and $\overline{BD}$ are the diagonals of rhombus ABCD. They intersect at point E. If line segment AE is 7 cm, find the measure of line segment $\overline{CE}$.

2. $\overline{AC}$ and $\overline{BD}$ are the diagonals of rhombus ABCD. They intersect at point E. If $m\overline{CE} = 4x$ and $m\overline{AE} = 52$ in, find the value of x.

3. $\overline{AC}$ and $\overline{BD}$ are diagonals of the rhombus ABCD that intersect at point E. If $m\overline{BE} = 4x - 10$ and $m\overline{ED} = x + 11$, find the value of x, the measure of line segment $\overline{BE}$, and the measure of diagonal $\overline{BD}$.

4. ∠AEB is created by two intersecting diagonals of the rhombus ABCD. If m□AEB = 5x, find the value of x.

LAB #36: THE PROPERTIES OF A TRAPEZOID

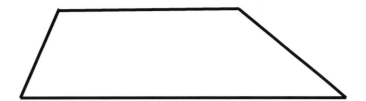

1. Select: [EDIT], [PREFERENCES], [TEXT], check the box [FOR ALL NEW POINTS] and click [OK].
2. Select: the Line Segment tool and construct a horizontal line segment.
3. Select: the Point tool and plot a point anywhere above line segment $\overline{AB}$.
4. Select: the Arrow tool, [EDIT], [SELECT ALL], [EDIT], [SELECT CHILDREN], [CONSTRUCT], [PARALLEL LINE], [CONSTRUCT] and [POINT ON PARALLEL LINE].
5. Click on point D, and keeping the left button depressed, drag it one inch to the left of point C and deselect.
6. Select: line $\overleftrightarrow{CD}$, [DISPLAY] and [HIDE PARALLEL LINE].
7. Select: points A, then D, then C, then B, [CONSTRUCT], [SEGMENTS], [MEASURE], [LENGTH] and deselect.
8. Select: the Text tool, double click in the blank region to open a dialog box and describe your observations about the measures of the sides of the trapezoid.
9. Select: the Arrow tool and click in the blank region to deselect.
10. Select: points A, then B, then C, [MEASURE], [ANGLE] and deselect.
11. Do the same to measure remaining angles $\angle$BCD, $\angle$CDA, and $\angle$DAB.
12. Select: [NUMBER], [CALCULATE], click on the caption that shows m$\angle$ABC, [+], click on the caption that shows m $\angle$BCD, [+], click on the caption that shows m$\angle$CDA, [+], click on the caption that shows m$\angle$DAB and [OK].

13. Click on any labeled point, and keeping the left button depressed, drag it. Observe the value that remains the same as you do so.
14. Select: the Text tool, double click in the blank region to open a dialog box and (1) describe your observations about the measures of the angles of trapezoid, (2) explain to what value the sum of the angles is equal.
15. Select: the Arrow tool, [NUMBER], [CALCULATE], click on the caption that shows m∠ABC, [+], click on the caption that shows m∠BCD and [OK].
16. Select: [NUMBER], [CALCULATE], click on the caption that shows m∠CDA, [+], click on the caption that shows m∠DAB and [OK].
17. Select: the Text tool, double click in the blank region to open a dialog box and (1) describe your observations about the sum of each pair of upper and lower consecutive angles on the non-parallel sides of the trapezoid, (2) name parallel sides of the trapezoid, (3) give your own definition of a trapezoid.

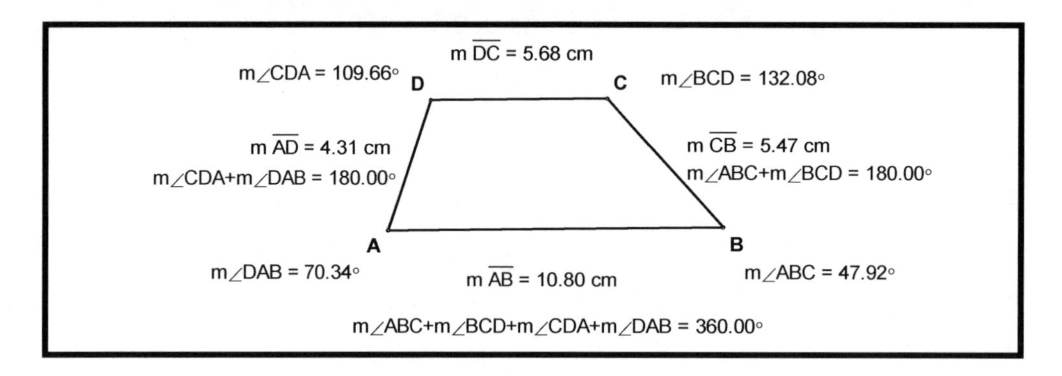

SUGGESTED EXERCISES

Type the solutions to the following problems in the blank region of your sketch.

1. ABCD is a trapezoid. If the two lower angles m∠A = 51° and m∠B = 73°, find the measures of two upper angles ∠C and ∠D.

2. ABCD is a trapezoid. If m∠A = 40°, and m∠D = x, find x.

3. ABCD is a trapezoid, with bases $\overline{AB}$ and $\overline{CD}$. If m∠B = x + 10, and m∠C = 3x + 40. Find x and m∠B and m∠C.

LAB #37: THE PROPERTIES OF THE ISOSCELES TRAPEZOID

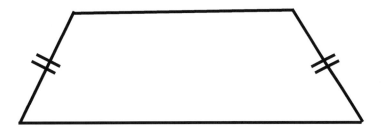

1. Select: [EDIT], [PREFERENCES], [TEXT], check the box [FOR ALL NEW POINTS] and click [OK].
2. Select: the Point tool and plot two points approximately one inch apart from each other in the middle of your screen.
3. Select: the Arrow tool and click in the blank region to deselect.
4. Select: point A, then B, [CONSTRUCT], [CIRCLE BY CENTER + POINT], [CONSTRUCT] and [POINT ON CIRCLE].
5. Click on point B, and keeping the left button depressed, drag it along the circle below and to the left of point A.
6. Click on point C, and keeping the left button depressed, drag it along the circle below and to the right of point A and deselect.
7. Select: the circle, [DISPLAY] and [HIDE CIRCLE].
8. Select: [EDIT], [SELECT ALL], [CONSTRUCT], [SEGMENTS] and deselect.
9. Select: line segments $\overline{AC}$, $\overline{AB}$, [CONSTRUCT], [MIDPOINTS] and deselect.
10. Select: the Arrow tool, sides $\overline{AB}$, $\overline{AC}$, point A, [DISPLAY] and [HIDE OBJECTS].
11. Select: points B, then E, then D, then C, [CONSTRUCT], [SEGMENTS], [MEASURE], [LENGTHS] and deselect.
12. Select: the Text tool, double click in the blank region to open a dialog box and write your observations about the measures of two nonparallel sides $\overline{BE}$ and $\overline{DC}$.
13. Select: the Arrow tool and click in the blank region to deselect.
14. Select: points C, then B, then E, [MEASURE], [ANGLE] and deselect.

15. Select: points B, then C, then D, [MEASURE] and [ANGLE].
16. Select: the Text tool, double click in the blank region to open a dialog box and describe your observations about the measures of the lower base angles $\angle CBE$ and $\angle BCD$.
17. Select: the Arrow tool and click in the blank region to deselect.
18. Select: points B, then E, then D, [MEASURE], [ANGLE] and deselect.
19. Select: points C, then D, then E, [MEASURE] and [ANGLE].
20. Select: the Text tool, double click in the blank region to open a dialog box and write your observations about the measure of the upper base angles $\angle BED$ and $\angle CDE$.
21. Select: the Arrow tool, click on point B or C, and keeping the left button depressed drag it. Observe what remains equal as you do so.
22. Select: the Text tool, double click in the blank region to open a dialog box and write your own definition of an isosceles trapezoid.

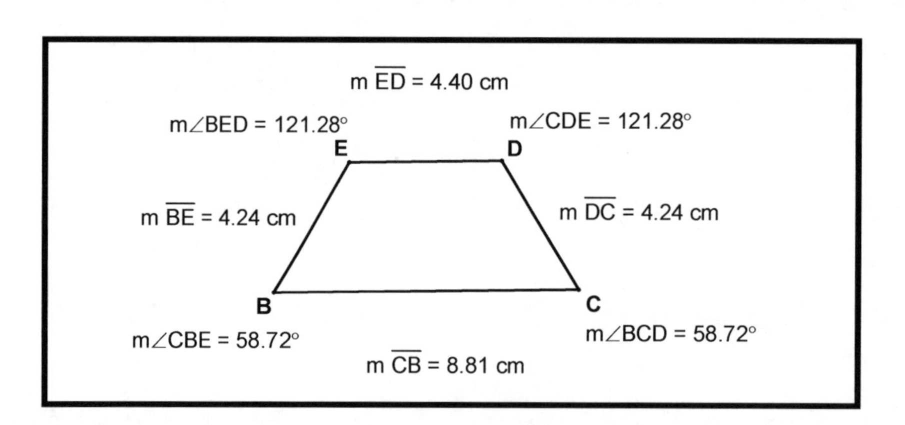

SUGGESTED EXERCISES

Type the solutions to the following problems in the blank region of your sketch.

1. In isosceles trapezoid BCDE, $\overline{BE} \cong \overline{CD}$. If $m\angle B = 48°$, find $m\angle C$.

2. In isosceles trapezoid BCDE, $\overline{BE} \cong \overline{CD}$. If $m\angle E = 120°$ and $m\angle D = 5x$, find the value of x.

3. $\overline{BE}$ and $\overline{CD}$ are the non-parallel sides of an isosceles trapezoid. If $m\overline{BE} = x + 3$, $m\overline{CD} = 2x + 2$, find the length of $\overline{BE}$.

LAB #38: THE DIAGONALS OF THE ISOSCELES TRAPEZOID

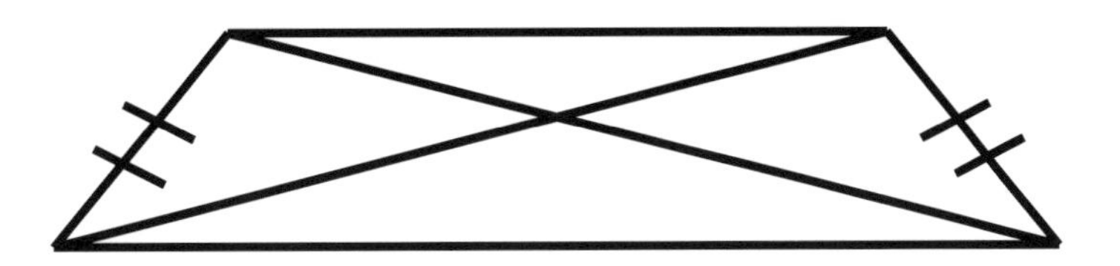

1. Select: [EDIT], [PREFERENCES], [TEXT], check the box [FOR ALL NEW POINTS] and click [OK].
2. Select: the Point tool and plot two points approximately one inch apart from each other in the middle of your screen.
3. Select: the Arrow tool and click in the blank region to deselect.
4. Select: points A, then B, [CIRCLE BY CENTER + POINT], [CONSTRUCT] and [POINT ON CIRCLE].
5. Click on point B, and keeping the left button depressed, drag it along the circle below and to the left of point A.
6. Click on point C, and keeping the left button depressed, drag it along the circle below and to the right of point A and deselect.
7. Select: the circle [DISPLAY] and [HIDE CIRCLE].
8. Select: [EDIT], [SELECT ALL], [CONSTRUCT], [SEGMENTS] and deselect.
9. Select: line segments $\overline{AC}$, $\overline{AB}$, [CONSTRUCT], [MIDPOINTS] and deselect.
10. Select: the Arrow tool, sides $\overline{AB}$, $\overline{AC}$, point A, [DISPLAY] and [HIDE OBJECTS].
11. Select: points B, then E, then D, then C, [CONSTRUCT], [SEGMENTS] and deselect.
12. Select: points B, then D, then C, then E, [CONSTRUCT], [SEGMENTS] and deselect.
13. Select: diagonals $\overline{CE}$, $\overline{BD}$, [CONSTRUCT], [INTERSECTION] and deselect.
14. Select: points C, then F, then E, [CONSTRUCT], [SEGMENTS], [MEASURE], [LENGTHS] and deselect.
15. Select: points B, then F, then D, [CONSTRUCT], [SEGMENTS], [MEASURE], [LENGTHS] and deselect.

16. Select: the Text tool, double click in the blank region to open a dialog box and (1) describe your observations about the measures of the diagonals $\overline{CE}$ and $\overline{BD}$, (2) describe your observations about the measures of line segments $\overline{BF}$ and $\overline{FD}$, $\overline{CF}$ and $\overline{FE}$, (3) explain if diagonals $\overline{BD}$ and $\overline{CE}$ bisect each other.
17. Select: the Arrow tool and click in the blank region to deselect.
18. Select: points B, then D, then E, [MEASURE], [ANGLE] and deselect.
19. Do the same to measure $\angle$BDC.
20. Select: the Text tool, double click in the blank region to open a dialog box and (1) describe your observations about the measures of the angles $\angle$BDE and $\angle$BDC, (2) explain if diagonal $\overline{BD}$ bisects $\angle$CDE (divides it in half).
21. Select: the Arrow tool, click on point B or C, and keeping the left button depressed, drag it. Observe the values that remain equal as you do so.
22. Select: the Text tool, double click in the blank region to open a dialog box and explain what you have learned about the diagonals of an isosceles trapezoid.

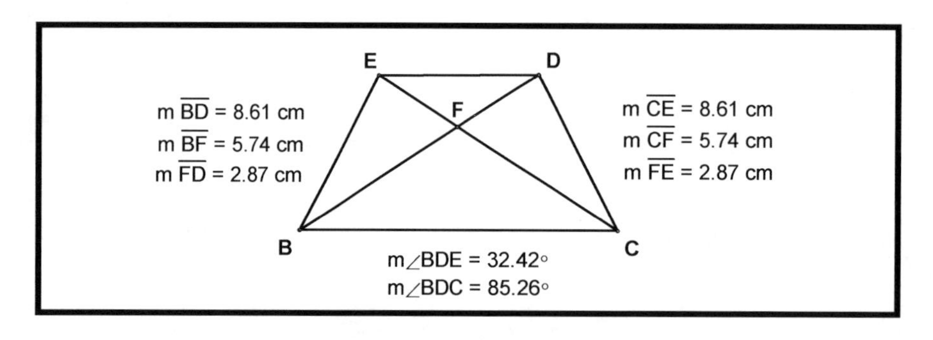

SUGGESTED EXERCISES

Type the solutions to the following problems in the blank region of your sketch.

1. $\overline{BD}$ and $\overline{CE}$ are the diagonals of isosceles trapezoid BCDE. If m$\overline{CE}$ = 12 cm, find the length of the diagonal $\overline{BD}$.

2. $\overline{BD}$ and $\overline{CE}$ are the diagonals of isosceles trapezoid BCDE. If m$\overline{CE}$ = 21 cm and m$\overline{BD}$= 3x, find the value of x.

3. $\overline{BD}$ and $\overline{CE}$ are the diagonals of isosceles trapezoid BCDE. If m$\overline{CE}$ = 3x + 8 and m$\overline{BD}$ = x + 12, find the length of $\overline{CE}$.

LAB #39: THE CORRESPONDING ANGLES OF SIMILAR TRIANGLES

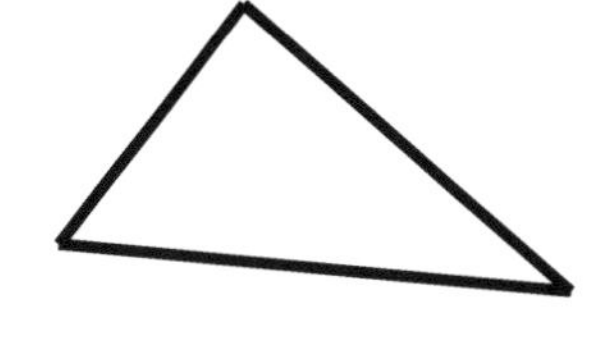

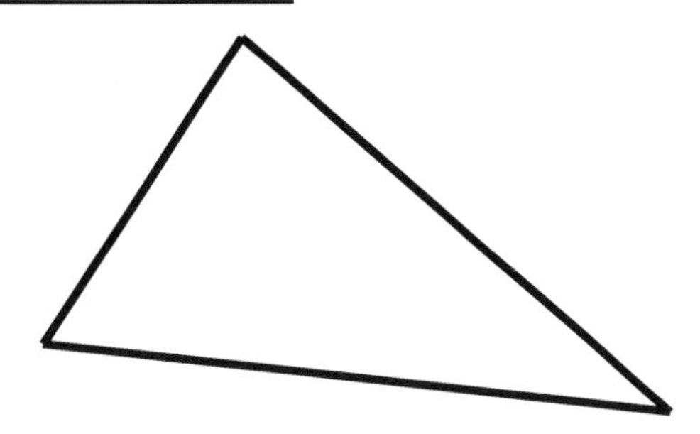

1. Select: [EDIT], [PREFERENCES], [TEXT], check the box [FOR ALL NEW POINTS] and click [OK].
2. Select: the Point tool and plot three points as the vertices of a triangle in the lower part of your screen.
3. Select: the Arrow tool, [EDIT], [SELECT ALL], [CONSTRUCT], [SEGMENTS] and deselect.
4. Select: line segment $\overline{AC}$, [CONSTRUCT], [POINT ON SEGMENT] and deselect.
5. Select: point D, line segment $\overline{BC}$, [CONSTRUCT], [PARALLEL LINE] and deselect.
6. Select: the parallel line passing through point D, line segment $\overline{AB}$, [CONSTRUCT], [INTERSECTION] and deselect.
7. Select: line $\overleftrightarrow{DE}$, [DISPLAY] and [HIDE PARALLEL LINE].
8. Select: points A, then E, then D, [TRANSFORM], [TRANSLATE], type [5] in the displayed window marked as [FIXED DISTANCE], [TRANSLATE], [CONSTRUCT], [SEGMENTS] and deselect.
9. Select: points D, E, [DISPLAY] and [HIDE POINTS].
10. Select: points A, then B, then C, [MEASURE], [ANGLE] and deselect.
11. Do the same to measure $\angle$ACB, $\angle$BAC, $\angle$A'E'D', $\angle$A'D'E', $\angle$D'A'E'.
12. Select: the Text tool, double click in the blank region to open a dialog box and (1) describe your observations about the measures of the corresponding angles: $\angle$ABC and $\angle$A'E'D', $\angle$ACB and $\angle$A'D'E', $\angle$BAC and $\angle$D'A'E',

    (2) state whether or not ΔABC and ΔA'E'D' have the same shape,

(3) given that similar triangles have the same shape, state whether or not ΔA'E'D' and ΔABC are similar.

13. Select: the Arrow tool, click on the point C or D', and keeping the left button depressed, drag it. Observe the measures that remain equal as you do so.
14. Select: the Text tool, double click in the blank region to open a dialog box and explain what makes ΔABC and ΔA'E'D' similar.

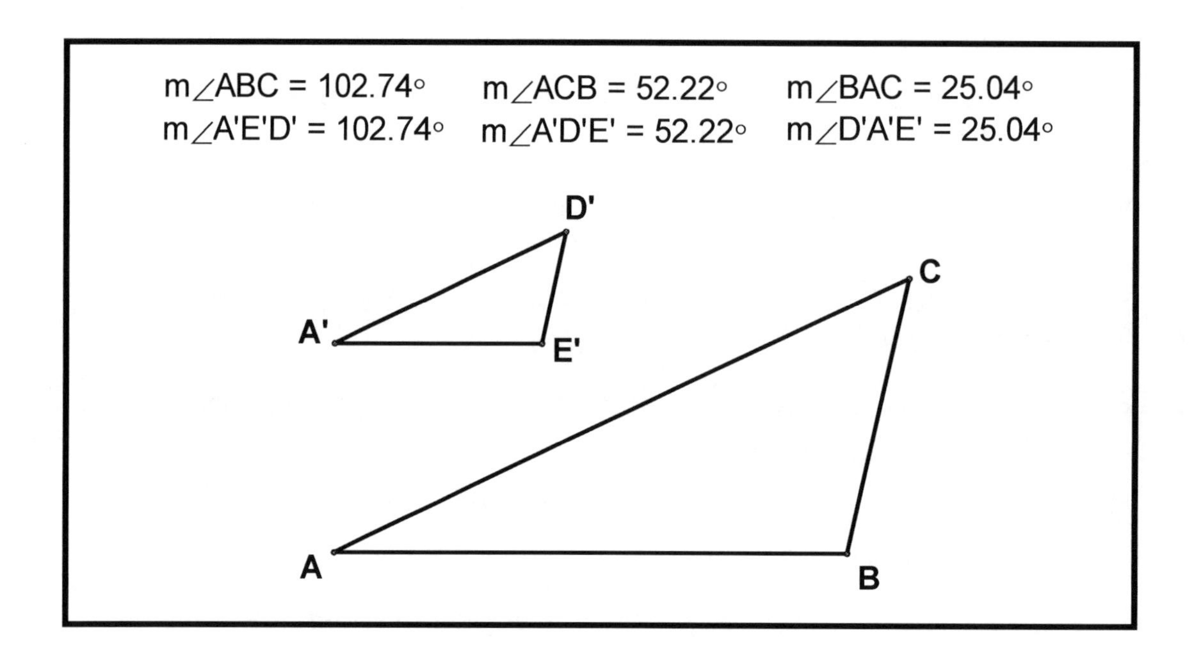

SUGGESTED EXERCISES

Type the solutions to the following problems in the blank region of your sketch.

1. In ΔABC m$\angle$A = 42°, m$\angle$B = 71°, and m$\angle$C = 67°. ΔA'E'D' is similar to ΔABC, find the degree measure of $\angle$A'E'D' and $\angle$A'D'E'.

2. $\angle$A'E'D' and $\angle$ABC are corrresponding angles of two similar triangles. Find the value of x if m$\angle$A'E'D' = 3x – 20 and m$\angle$ABC = 70°.

3. $\angle$A'E'D' and $\angle$ABC are two corrresponding angles of two similar triangles, if m$\angle$A'E'D' = 2x + 20 and m$\angle$ABC = x + 35 , find the value of x and the measures of these two angles.

LAB #40: THE CORRESPONDING SIDES OF SIMILAR TRIANGLES

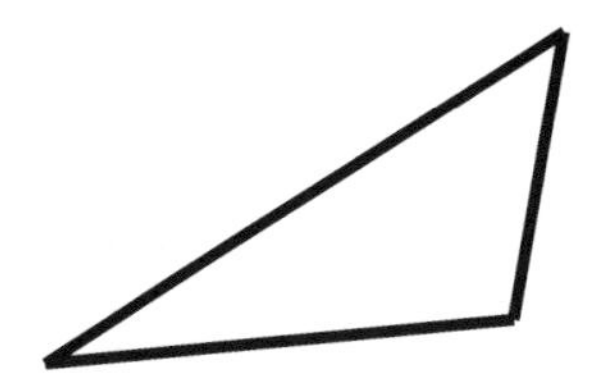

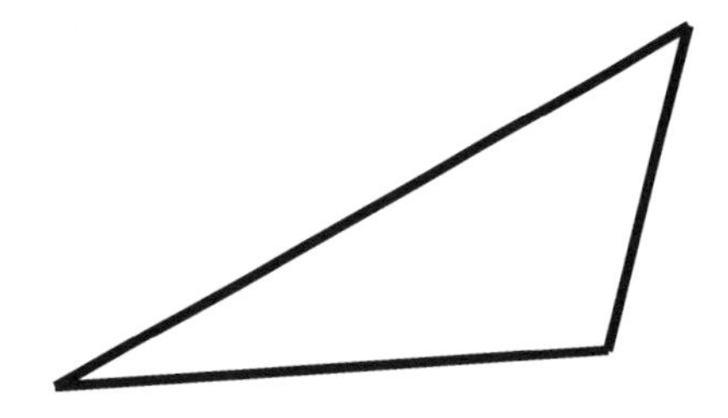

1. Select: [EDIT], [PREFERENCES], [TEXT], check the box [FOR ALL NEW POINTS] and click [OK].
2. Select: the Point tool and plot three points as the vertices of a triangle in the lower part of your screen.
3. Select: the Arrow tool, [EDIT], [SELECT ALL], [CONSTRUCT], [SEGMENTS], [MEASURE], [LENGTHS] and deselect.
4. Select: line segment $\overline{AC}$, [CONSTRUCT], [POINT ON SEGMENT] and deselect.
5. Select: point D, line segment $\overline{BC}$, [CONSTRUCT], [PARALLEL LINE] and deselect.
6. Select: the parallel line passing through point D, line segment $\overline{AB}$, [CONSTRUCT], [INTERSECTION] and deselect.
7. Select: line $\overleftrightarrow{DE}$, [DISPLAY] and [HIDE PARALLEL LINE].
8. Select: points A, then E, then D, [TRANSFORM], [TRANSLATE], type [5] in the displayed window marked as [FIXED DISTANCE] [TRANSLATE], [CONSTRUCT], [SEGMENTS], [MEASURE], [LENGTHS] and deselect.
9. Select: points D, E, [DISPLAY] and [HIDE POINTS].
10. Select: [NUMBER], [CALCULATE], click on the caption that shows $m\overline{AB}$, [÷], click on the caption that shows $m\overline{A'E'}$, [OK] and deselect.
11. Select: [NUMBER], [CALCULATE], click on the caption that shows $m\overline{BC}$, [÷], click on the caption that shows $m\overline{E'D'}$, [OK] and deselect.
12. Select: [NUMBER], [CALCULATE], click on the caption that shows $m\overline{CA}$, [÷], click on the caption that shows $m\overline{D'A'}$, [OK] and deselect.

13. Click on the point C, or D, and keeping the left button depressed, drag it. Observe the values that remain equal as you do so.
14. Select: the Text tool, double click in the blank region to open a dialog box, and (1) state whether or not ΔABC and ΔA'E'D' are equal in size,

    (2) explain what you learned about the ratios of the corresponding sides of two similar triangles,
    (3) explain what the value of the ratio of the corresponding sides tells you about the size of ΔABC compared to the size of ΔA'E'D',
    (4) write a proportion involving any two ratios of the corresponding sides.

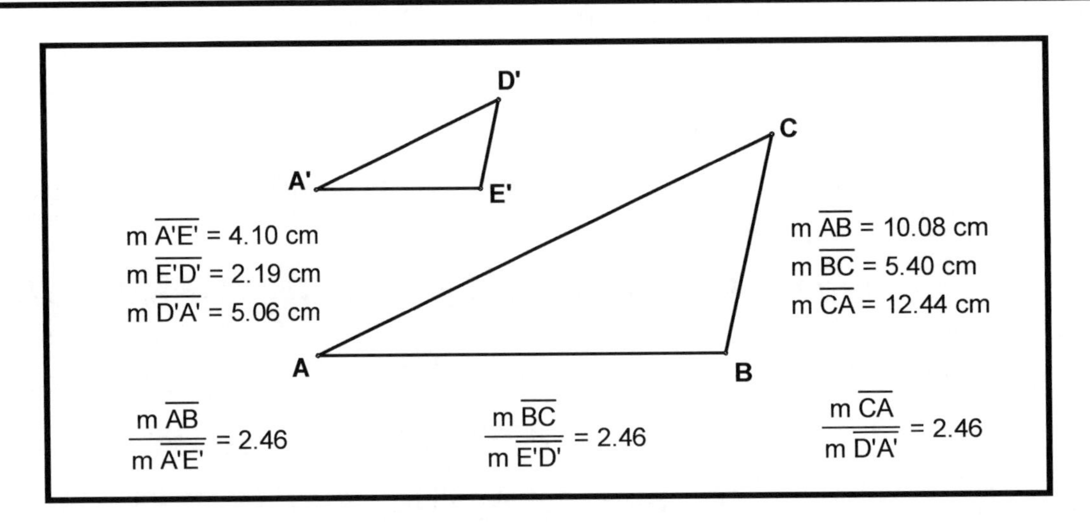

SUGGESTED EXERCISES

Type the solutions to the following problems in the blank region of your sketch.

1. ΔABC is similar to ΔA'E'D'. $m\overline{AC}$ = 12 in, $m\overline{AB}$ = 15 in and $m\overline{A'D'}$ = 8 in. Find $m\overline{A'E'}$.

2. The sides of a triangle measure 2, 7, and 8 inches. If the longest side of a similar triangle measures 16 inches, find the shortest side of this triangle.

3. The sides of a triangle measure 12, 16, and 24 inches. If the shortest side of a similar triangle measures 6 inches, find the longest side of this triangle.

LAB #41: THE CORRESPONDING PERIMETERS OF SIMILAR TRIANGLES

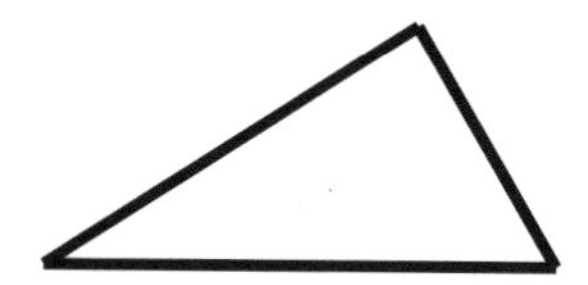

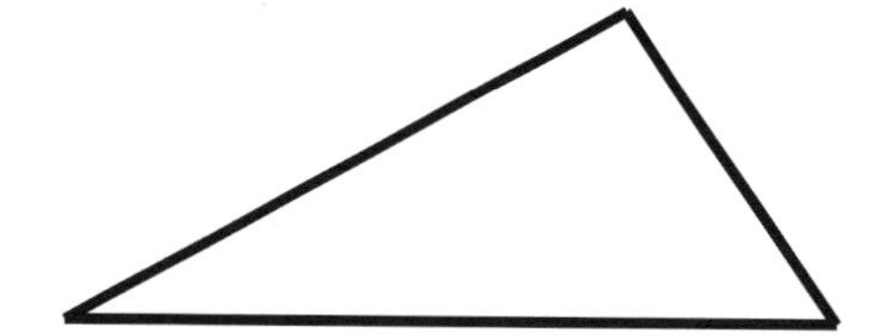

1. Select: [EDIT], [PREFERENCES], [TEXT], check the box [FOR ALL NEW POINTS] and click [OK].
2. Select: the Point tool and plot three points as the vertices of a triangle in the lower part of your screen.
3. Select: the Arrow tool, [EDIT], [SELECT ALL], [CONSTRUCT], [SEGMENTS], [MEASURE], [LENGTHS] and deselect.
4. Select: [NUMBER], [CALCULATE], click on the caption that shows $m\overline{AB}$, [+], click on the caption that shows $m\overline{BC}$, [+], click on the caption that shows $m\overline{CA}$, [OK] and deselect.
5. Select: line segment $\overline{AC}$, [CONSTRUCT], [POINT ON SEGMENT] and deselect.
6. Select: point D, line segment $\overline{BC}$, [CONSTRUCT], [PARALLEL LINE] and deselect.
7. Select: the parallel line passing through point D, line segment $\overline{AB}$, [CONSTRUCT], [INTERSECTION] and deselect.
8. Select: line $\overleftrightarrow{DE}$, [DISPLAY] and [HIDE PARALLEL LINE].
9. Select: points A, then E, then D, [TRANSFORM], [TRANSLATE], type [5] in the displayed window marked as [FIXED DISTANCE], [TRANSLATE], [CONSTRUCT], [SEGMENTS], [MEASURE], [LENGTHS] and deselect.
10. Select: points D, E, [DISPLAY] and [HIDE POINTS].
11. Select: [NUMBER], [CALCULATE], click on the caption that shows $m\overline{A'E'}$, [+], click on the caption that shows $m\overline{E'D'}$, [+], click on the caption that shows $m\overline{D'A'}$, [OK] and deselect.
12. Select: [NUMBER], [CALCULATE], click on the caption that shows $m\overline{AB}$, [÷], click on the caption that shows m $\overline{A'E'}$, [OK] and deselect.

13. Select: [NUMBER], [CALCULATE], click on the caption that shows the perimeter of ΔABC ($m\overline{AB} + m\overline{BC} + m\overline{AC}$), [÷], click on the caption that shows the perimeter of $\Delta A'E'D'$ ($m\overline{A'E'} + m\overline{D'E'} + m\overline{A'E'}$) and [OK].
14. Click on the point D', and keeping the left button depressed, drag it. Do the same with the point C. Observe what values remain equal to each other as you do so.
15. Select: the Text tool, double click in the blank region to open a dialog box and (1) explain what you learned about the ratios of the corresponding sides and the ratio of the corresponding perimeters of two similar triangles, 2) write a proportion involving these ratios.

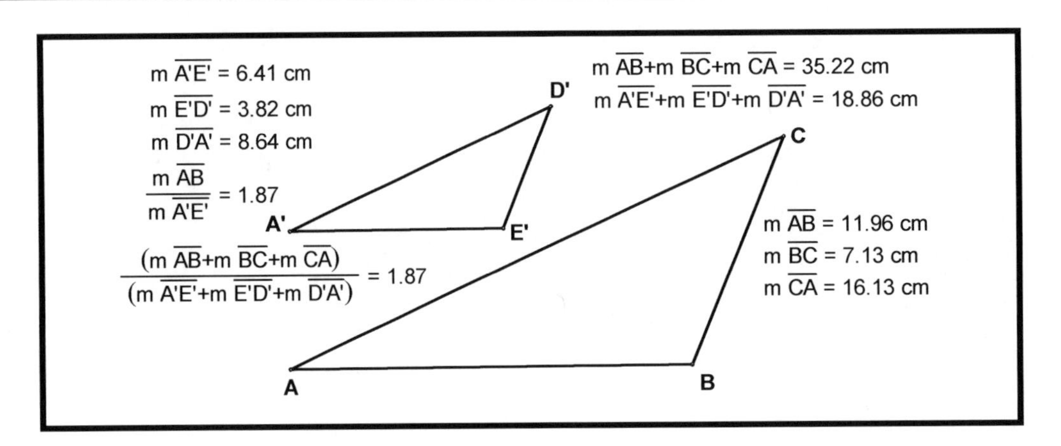

SUGGESTED EXERCISES

Type the solutions to the following problems in the blank region of your sketch.

1. If the ratio of the corresponding sides of two similar triangles is $\frac{3}{4}$, find the ratio of their corresponding perimeters.
2. The sides of a triangle measure 6, 8, and 10, respectively. What is the length of the shortest side of a similar triangle whose perimeter is 12?
3. The sides of a triangle measure respectively 3, 7, and 8. What is the length of the longest side of a similar triangle whose perimeter is 54?

LAB #42: THE CORRESPONDING AREAS OF SIMILAR TRIANGLES

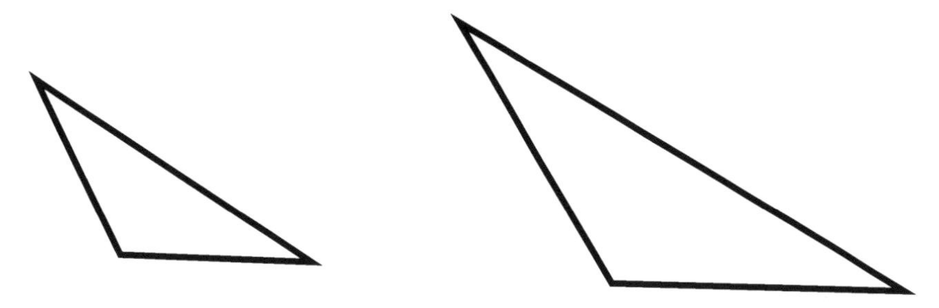

1. Select: [EDIT], [PREFERENCES], [TEXT], check the box [FOR ALL NEW POINTS] and click [OK].
2. Select: the Point tool and plot three points as the vertices of the triangle in the lower part of your screen.
3. Select: the Arrow tool, [EDIT], [SELECT ALL], [CONSTRUCT], [SEGMENTS], [MEASURE], [LENGTHS] and deselect.
4. Select: line segment $\overline{AC}$, [CONSTRUCT], [POINT ON SEGMENT] and deselect.
5. Select: point D, line segment $\overline{BC}$, [CONSTRUCT], [PARALLEL LINE] and deselect.
6. Select: the parallel line passing through point D, line segment $\overline{AB}$, [CONSTRUCT], [INTERSECTION] and deselect.
7. Select: line $\overleftrightarrow{DE}$, [DISPLAY] and [HIDE PARALLEL LINE].
8. Select: points A, then E, then D, [TRANSFORM], [TRANSLATE], type [5] in the displayed window marked as [FIXED DISTANCE], [TRANSLATE], [CONSTRUCT], [SEGMENTS], [MEASURE], [LENGTHS] and deselect.
9. Select: points D, E, [DISPLAY] and [HIDE POINTS].
10. Select: points A, B, C, [CONSTRUCT], [TRIANGLE INTERIOR], [MEASURE], [AREA] and deselect.
11. Select: points A', E', D', [CONSTRUCT], [TRIANGLE INTERIOR], [MEASURE], [AREA] and deselect.
12. Select: [NUMBER], [CALCULATE], [( ], click on the caption that shows $m\overline{AB}$, [÷], click on the caption that shows $m\overline{A'E'}$, [ )], [∧], [2], [OK] and deselect.

13. Select: [NUMBER], [CALCULATE], click on the caption that shows the measure of the area of ΔABC, [$\div$], click on the caption that shows the measure of the area of ΔA'E'D' and [OK].
14. Select: the Text Tool, double click in the blank region to open a dialog box and describe your observations about the square of the ratio of $\frac{AB}{A'E'}$ and the value of the ratio of $\frac{\text{Area}\Delta ABC}{\text{Area}\Delta A'E'D'}$.
15. Select: the Arrow tool, click on point D', and keeping the left button depressed, drag it. Do the same with the point C. Observe the values that remain equal as you do so.
16. Select: the Text tool, double click in the blank region to open a dialog box and (1) explain what you have learned about the square of the ratio of the corresponding sides and the ratio of the areas of two similar triangles, (2) write a proportion involving two ratios.

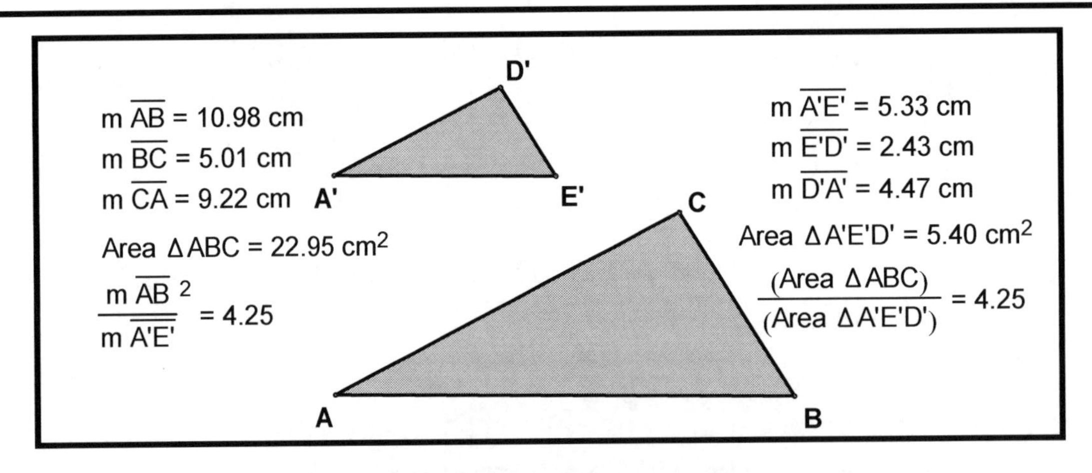

SUGGESTED EXERCISES

Type the solutions to the following problems in the blank region of your sketch.

1. The ratio of two corresponding sides of two similar triangles is $\frac{2}{3}$. Find the ratio of their areas.
2. The ratio of the areas of two similar triangles is $\frac{25}{16}$. Find the ratio of their corresponding sides.
3. The lengths of a pair of corresponding sides of two similar triangles are 4 inches and 6 inches. If the area of the first triangle is 20 square inches, find the area of the second triangle.

LAB #43: CONGRUENT TRIANGLES: S.S.S. ≅ S.S.S.

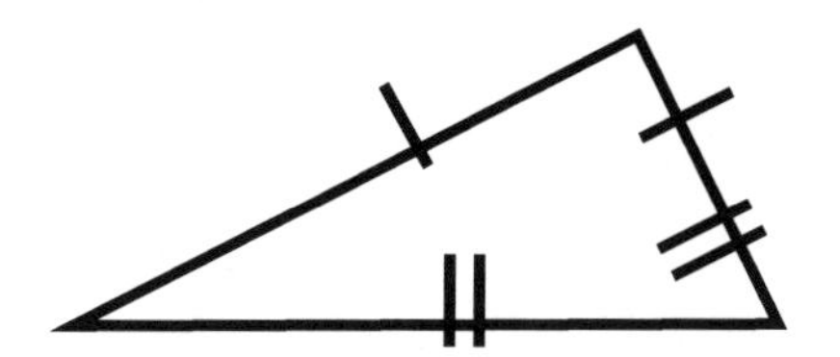

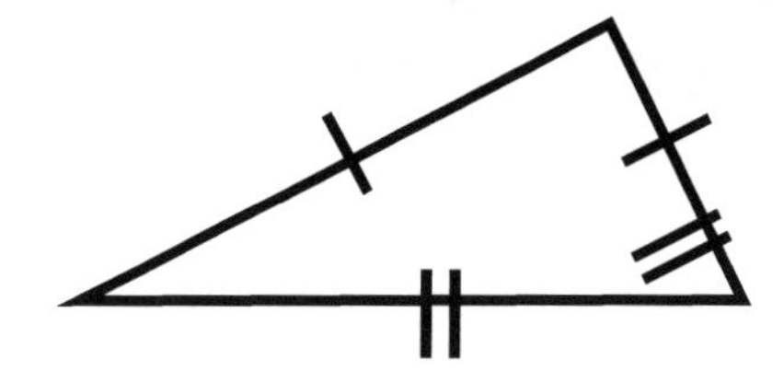

1. Select: [EDIT], [PREFERENCES], [TEXT], check the box [FOR ALL NEW POINTS] and click [OK].
2. Select: the Point tool and plot points A and B from left to right approximately three inches apart in the left part of your screen.
2. Plot point C anywhere above points A and B.
3. Select: the Arrow tool, [EDIT], [SELECT ALL], [CONSTRUCT], [SEGMENTS], [MEASURE], [LENGTHS] and deselect.
4. Select: points A, then B, then C, [MEASURE], [ANGLE] and deselect.
5. Do the same to measure ∠ACB and ∠BAC and deselect.
6. Select: the Point tool and plot a point approximately one inch to the right of point B.
7. Select: the Arrow tool and click in the blank region to deselect.
8. Select: points A, then D, [TRANSFORM] and [MARK VECTOR].
9. Select: points A, then B, then C, [TRANSFORM], [TRANSLATE], [TRANSLATE] and deselect.
10. Select: the Point tool and plot a point approximately one inch to the right of point B′.
11. Select: the Arrow tool and click in the blank region to deselect.
12. Select: points E, then B′, [EDIT], [ACTION BUTTONS], [MOVEMENT], [MEDIUM], [SLOW], [LABEL], type [CONGRUENT TRIANGLES], [OK] and deselect.
13. Select: point B′, [DISPLAY] and [HIDE POINT].
14. Select: point C′, [EDIT], [PROPERTIES], type a capital F in the display region, [OK] and deselect.
15. Select: points D, then E, then F, [CONSTRUCT], [SEGMENTS], [MEASURE], [LENGTHS] and deselect.
16. Select: points D, then E, then F, [MEASURE], [ANGLE] and deselect.
17. Do the same to measure ∠DFE and ∠EDF and deselect.
18. Select: the Text tool, double click in the blank region to open a dialog box and explain whether or not all sides and all angles of Δ ABC are equal in measure and, therefore congruent to all sides and all angles of Δ DEF.

19. Select: the Arrow tool and click in the blank region to deselect.
20. Click on the action button [CONGRUENT TRIANGLES] and observe the changes in the measures of the sides and the angles of Δ DEF.
21. Click on point C, and keeping the left button depressed, drag it to any new location on the plane and deselect.
22. Click on point D, and keeping the left button depressed, drag it to any new location on the plane and deselect.
23. Click on the action button [CONGRUENT TRIANGLES] and continue your observations of the changes in the measures of the sides and the angles of Δ DEF.
24. Select: the Text tool, double click in the blank region to open a dialog box and (1) explain what happens to the corresponding sides and the corresponding angles of Δ ABC and Δ DEF after they become congruent triangles (equal in size and shape), (2) explain what must be true about the corresponding sides and the corresponding angles of any two triangles to be called congruent.

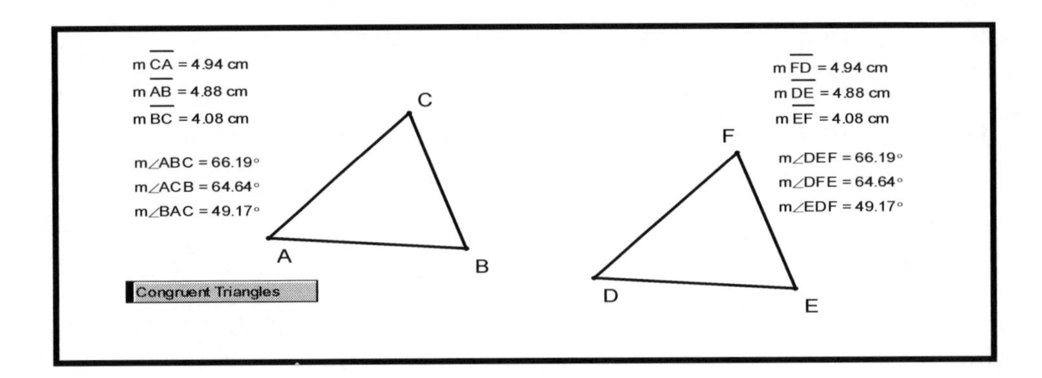

SUGGESTED EXERCISES

Type the solutions to the following problem in the blank region of your sketch.

1. $\Delta ABC \cong \Delta DEF$. If m$\overline{AB}$ = 7 cm, find the length of the corresponding side $\overline{DE}$ in ΔDEF.

2. $\Delta ABC \cong \Delta DEF$. If m$\overline{AC}$ = 3x and m$\overline{FD}$ = 42 cm, find the value of x.

3. $\Delta ABC \cong \Delta DEF$. If m$\overline{AB}$ = 3x and m$\overline{DE}$ = 2x + 10, find m$\overline{AB}$ and m$\overline{DE}$.

4. Nicole drew two triangles whose corresponding angles measured 50, 60, and 70 degrees, respectively. The triangles were not congruent. Explain why this could have happened.

LAB #44: CONGRUENT TRIANGLES: S.A.S. ≅ S.A.S.

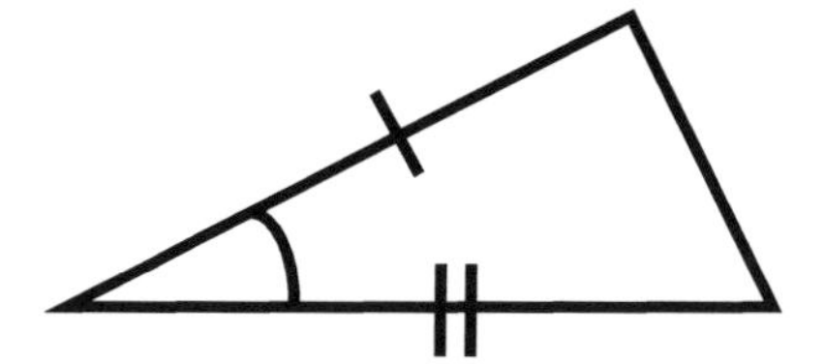
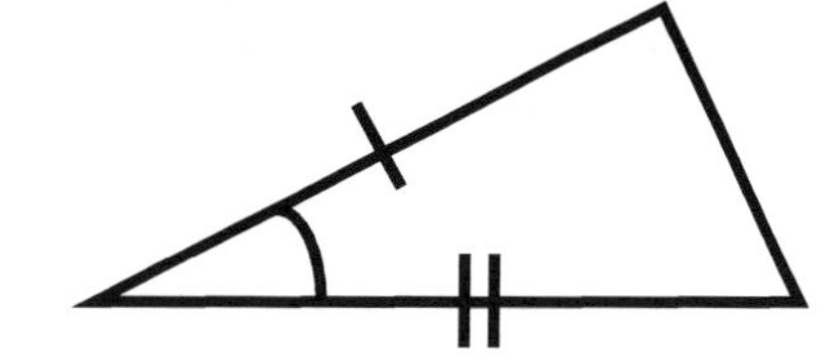

1. Select: [EDIT], [PREFERENCES], [TEXT], check the box [FOR ALL NEW POINTS] and click [OK].
2. Select: the Point tool and plot points A and B from left to right approximately three inches apart in the left part of your screen.
2. Plot point C anywhere above points A and B.
3. Select: the Arrow tool, [EDIT], [SELECT ALL], [CONSTRUCT], [SEGMENTS], [MEASURE], [LENGTHS] and deselect.
4. Select: points B, then A, then C, [MEASURE], [ANGLE] and deselect.
5. Select: the Point tool and plot a point approximately one inch to the right of point B.
6. Select: the Arrow tool and click in the blank region to deselect.
7. Select: points A, then D, [TRANSFORM] and [MARK VECTOR].
8. Select: points A, then B, then C, [TRANSFORM], [TRANSLATE], [TRANSLATE] and deselect.
9. Select: the Point tool and plot a point approximately one inch to the right of point B′.
10. Select: the Arrow tool and click in the blank region to deselect.
11. Select: points E, then B′, [EDIT], [ACTION BUTTONS], [MOVEMENT], [MEDIUM], [SLOW], [LABEL], type [MOVE E], [OK] and deselect.
12. Select: point B′, [DISPLAY] and [HIDE POINT].
13. Select: point C′, [EDIT], [PROPERTIES], type a capital F in the display region, [OK] and deselect.
14. Select: points D, then E, then F, [CONSTRUCT], [SEGMENTS], [MEASURE], [LENGTHS] and deselect.
15. Select: points E, then D, then F, [MEASURE], [ANGLE] and deselect.
16. Select: the captions that show $m\overline{BC}$, $m\overline{EF}$, [EDIT], [ACTION BUTTONS], [HIDE/SHOW] and deselect.
17. Click on the action button [HIDE DISTANCE MEASUREMENTS].
18. Select: the Text tool, double click in the blank region to open a dialog box and explain whether or not two sides and the included angle of ΔABC are equal in measure and, therefore, congruent to two sides and the included angle of ΔDEF.
19. Select: the Arrow tool and click in the blank region to deselect.
20. Click on the action button [MOVE E] and observe the changes in the measures of the two sides and the included angle of ΔDEF.

21. Click on point C, and keeping the left button depressed, drag it to any new location on the plane and deselect.
22. Click on point D, and keeping the left button depressed, drag it to any new location on the plane and deselect.
23. Click on the action button [MOVE E] and continue your observations of the changes in the measures of the two sides and the included angle of ΔDEF.
24. Select: the Text tool, double click in the blank region to open a dialog box and explain whether or not ΔABC and ΔDEF appear to become congruent (equal in size and shape) after two sides and the included angle of one triangle become congruent to two sides and the included angle of the second triangle.
25. Select: the Arrow tool and click in the blank region to deselect.
26. Click on the action button [SHOW DISTANCE MEASUREMENTS], deselect and observe $m\overline{BC}$ and $m\overline{EF}$.
27. Select: the Text tool, double click in the blank region to open a dialog box and (1) use your measurements of sides $\overline{BC}$ and $\overline{EF}$ to justify your assumption of whether or not $\Delta ABC \cong \Delta DEF$ by the S.S.S. $\cong$ S.S.S. rule, (2) explain what must be true about two sides and the <u>included</u> angle (S.A.S.) of two triangles in order to be called congruent.

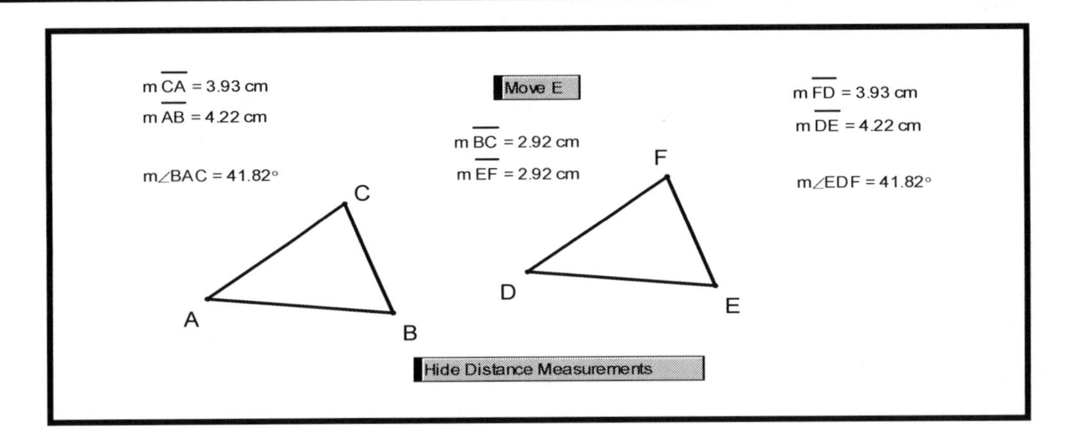

SUGGESTED EXERCISES

Type the solutions to the following problems in the blank region of your sketch.

1. $\Delta ABC \cong \Delta DEF$. If $m\overline{AC} = 4$ cm, $m\angle BAC = 36°$, and $m\overline{AB} = 7$ cm, find $m\overline{FD}$, $m\angle EDF$, and $m\overline{DE}$.

2. $\Delta ABC \cong \Delta DEF$. If $m\angle A = 4x$ and $m\angle D = 72°$, find the value of x.

3. $\Delta ABC \cong \Delta DEF$. If $m\overline{AB} = 2x + 5$ and $m\overline{DE} = 3x + 4$, find $m\overline{AB}$ and $m\overline{DE}$.

4. Gordon drew two triangles in which two sides and an angle measured 5 cm, 8 cm, and 60 degrees, respectively. The triangles were not congruent. Explain why this could have happened.

LAB #45: CONGRUENT TRIANGLES: A.S.A. ≅ A.S.A.

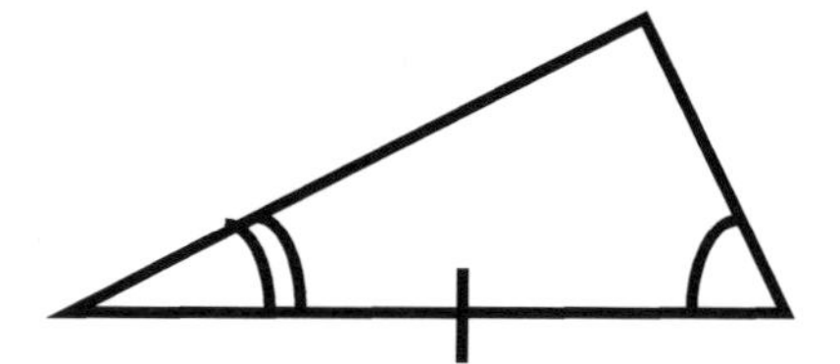

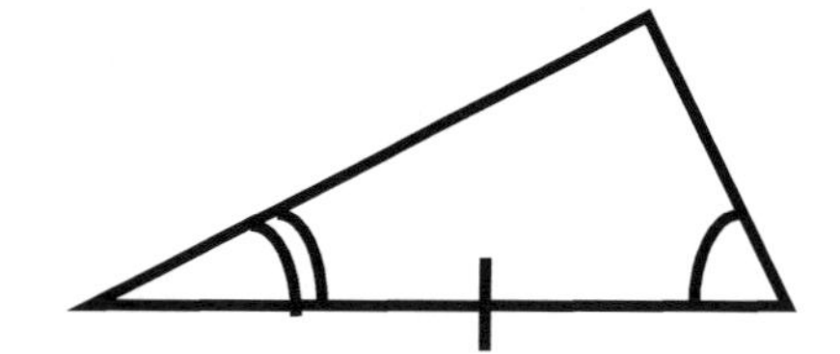

1. Select: [EDIT], [PREFERENCES], [TEXT], check the box [FOR ALL NEW POINTS] and click [OK].
2. Select: the Point tool and plot points A and B from left to right approximately three inches apart in the left part of your screen.
2. Plot point C anywhere above points A and B.
3. Select: the Arrow tool, [EDIT], [SELECT ALL], [CONSTRUCT], [SEGMENTS], [MEASURE], [LENGTHS] and deselect.
4. Select: points B, then A, then C, [MEASURE], [ANGLE] and deselect.
5. Select: points A, then B, then C, [MEASURE], [ANGLE] and deselect.
6. Select: the Point tool and plot a point approximately one inch to the right of point B.
7. Select: the Arrow tool and click in the blank region to deselect.
8. Select: points A, then D, [TRANSFORM] and [MARK VECTOR].
9. Select: points A, then B, then C, [TRANSFORM], [TRANSLATE], [TRANSLATE] and deselect.
10. Select: the Point tool and plot a point approximately one inch to the right of point B′.
11. Select: the Arrow tool and click in the blank region to deselect.
12. Select: points E, then B′, [EDIT], [ACTION BUTTONS], [MOVEMENT], [MEDIUM], [SLOW], [LABEL], type [MOVE E], [OK] and deselect.
13. Select: point B′, [DISPLAY] and [HIDE POINT].
14. Select: point C′, [EDIT], [PROPERTIES], type a capital F in the display region, [OK] and deselect.
15. Select: points D, then E, then F, [CONSTRUCT], [SEGMENTS], [MEASURE], [LENGTHS] and deselect.
16. Select: points E, then D, then F, [MEASURE], [ANGLE] and deselect.
17. Select: points D, then E, then F, [MEASURE], [ANGLE] and deselect.
18. Select: the captions that show $m\overline{BC}$, $m\overline{CA}$, $m\overline{EF}$, $m\overline{FD}$, [EDIT], [ACTION BUTTONS], [HIDE/SHOW] and deselect.
19. Click on the action button [HIDE DISTANCE MEASUREMENTS].
20. Select: the Text tool, double click in the blank region to open a dialog box and explain whether or not two angles and the included side of ΔABC are equal in measure and, therefore, congruent to two angles and the included side of ΔDEF.
21. Select: the Arrow tool and click in the blank region to deselect.

22. Click on the action button [MOVE E] and observe the changes in the measures of the two angles and the included side of Δ DEF.
23. Click on point C, and keeping the left button depressed, drag it to any new location on the plane and deselect.
24. Click on point D, and keeping the left button depressed, drag it to any new location on the plane and deselect.
25. Click on the action button [MOVE E] and continue your observations of the changes in the measures of the two angles and the included side of Δ DEF.
26. Select: the Text tool, double click in the blank region to open a dialog box and explain whether or not Δ ABC and Δ DEF appear to become congruent (equal in size and shape) after two angles and the included side of one triangle become congruent to two angles and the included side of the second triangle.
27. Select: the Arrow tool and click in the blank region to deselect.
28. Click on the action button [SHOW DISTANCE MEASUREMENTS], deselect and observe $m\overline{BC}$, $m\overline{CA}$, $m\overline{EF}$, $m\overline{FD}$.
29. Select: the Text tool, double click in the blank region to open a dialog box and (1) use your measurements of sides $\overline{BC}$, $\overline{CA}$, $\overline{EF}$, $\overline{FD}$ to justify your assumption of whether or not $\Delta ABC \cong \Delta DEF$ by the S.S.S. $\cong$ S.S.S. rule,
(2) explain what must be true about two angles and the <u>included</u> side (A.S.A.) of two triangles in order to be called congruent.

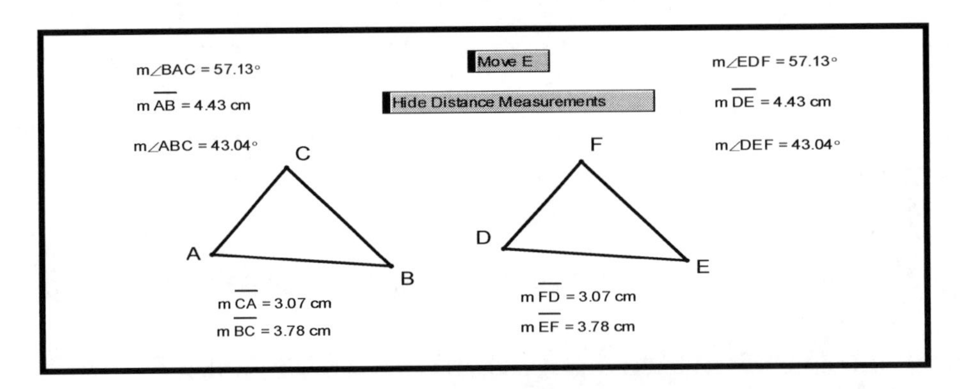

SUGGESTED EXERCISES

Type the solutions to the following problems in the blank region of your sketch.

1. $\Delta ABC \cong \Delta DEF$. If $m\angle BAC = 56°$, $m\overline{AB} = 5$ in, and $m\angle ABC = 61°$, find $m\angle EDF$, $m\overline{DE}$, and $m\angle DEF$.

2. $\Delta ABC \cong \Delta DEF$. If $m\angle B = 6x$ and $m\angle E = 48°$, find the value of x.

3. $\Delta ABD \cong \Delta DEF$. If $m\overline{AB} = 8x - 6$ and $m\overline{DE} = 5x + 12$, find $m\overline{AB}$ and $m\overline{DE}$.

LAB #46: A REFLECTION IN A LINE

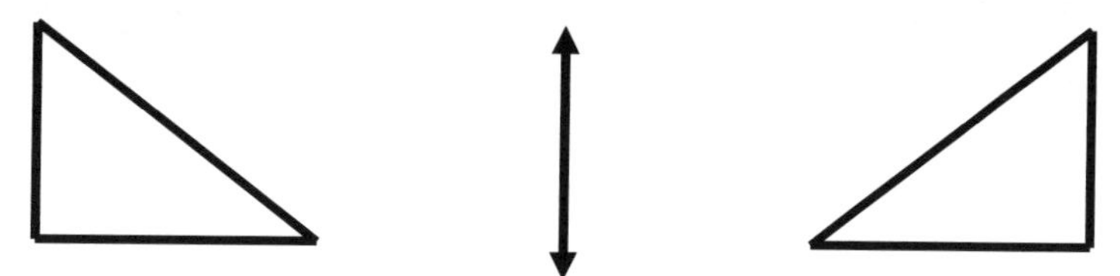

1. Select: [EDIT], [PREFERENCES], [TEXT], check the box [FOR ALL NEW POINTS] and click [OK].
2. Select: the Point tool and plot two points in the middle of your screen as the endpoints of a vertical line segment.
3. Select: the Arrow tool, [EDIT], [SELECT ALL], [CONSTRUCT], [LINE] and deselect.
4. Select: the Point tool and randomly plot 3 points to the left of line $\overleftrightarrow{AB}$ as the vertices of a triangle.
5. Select: the Arrow tool and click in the blank region to deselect.
6. Select: points C, then D, then E, [CONSTRUCT], [SEGMENTS], [MEASURE], [LENGTHS] and deselect.
7. Select: points C, then D, then E, [MEASURE], [ANGLE] and deselect.
8. Do the same to measure $\angle CED$ and $\angle DCE$.
9. Select: line of reflection $\overleftrightarrow{AB}$, [TRANSFORM] and [MARK MIRROR].
10. Select: points C, then D, then E, [TRANSFORM], [REFLECT], [CONSTRUCT], [SEGMENTS], [MEASURE], [LENGTHS] and deselect.
11. Select: the Text tool, double click in the blank region to open a dialog box, and describe your observations about the measures of sides $\overline{CD}$ and $\overline{C'D'}$, $\overline{DE}$ and $\overline{D'E'}$, $\overline{CE}$ and $\overline{C'E'}$.
12. Select: the Arrow tool and click in the blank region to deselect.
13. Select: points C', then D', then E', [MEASURE], [ANGLE] and deselect.
14. Do the same to measure $\angle C'E'D'$ and $\angle D'C'E'$.
15. Select: the Text tool, double click in the blank region to open a dialog box, and describe your observations about the measures of $\angle CDE$ and $\angle C'D'E'$, $\angle CED$ and $\angle C'E'D'$, $\angle DCE$ and $\angle D'C'E'$.
16. Select: the Arrow tool, click on any labeled point, and keeping the left button depressed, drag it. Observe what angles and line segments remain equal as you do so.
17. Select: the Text tool, double click in the blank region to open a dialog box, and describe what happens to the length of the line segments and angle measures of a polygon under a reflection in a line.

18. Select: the Arrow tool and click in the blank region to deselect.
19. Select: points C, then D, then E, [CONSTRUCT], [TRIANGLE INTERIOR], [CONSTRUCT], [POINT ON PERIMETER], [TRANSFORM], [REFLECT] and deselect.
20. Select: point F, line $\overleftrightarrow{AB}$, [MEASURE], [DISTANCE] and deselect.
21. Select: point F', line $\overleftrightarrow{AB}$, [MEASURE], [DISTANCE] and deselect.
22. Select: points F, F', [CONSTRUCT], [SEGMENT], [DISPLAY], [LINE STYLE], [THIN], [DISPLAY], [LINE STYLE], [DASHED] and deselect.
23. Select: point F, [EDIT], [ACTION BUTTONS], [ANIMATION], [MEDIUM], [SLOW] and [OK].
24. Click on [ANIMATE POINT] and observe the distances that remain equal as points F and F' move along the sides of Δ CDE and Δ C'D'E'.
25. Click on [ANIMATE POINT] again to stop the animation.
26. Select: the Text tool, double click in the blank region to open a dialog box and write what you learned about the distance from any point of the original polygon and its image about the line of reflection.

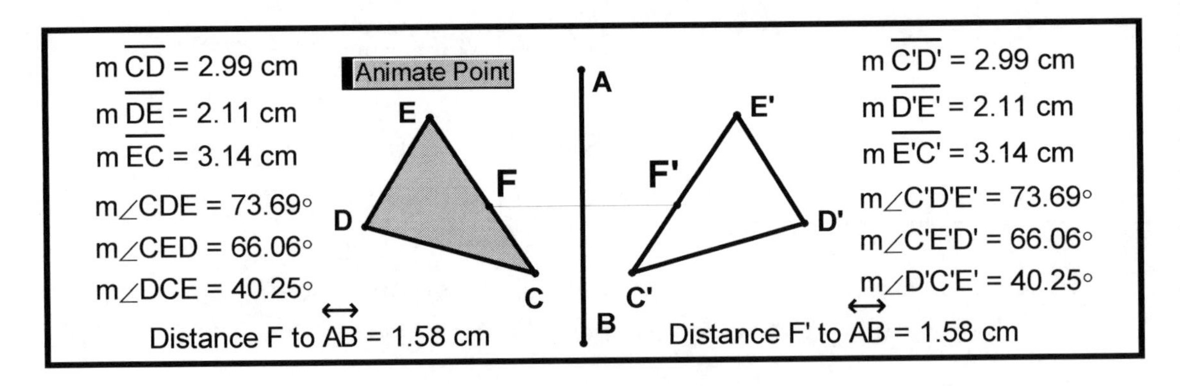

SUGGESTED EXERCISES

Type the solutions to the following problems in the blank region of your sketch.

1. Δ CDE is reflected over the line of reflection, $\overleftrightarrow{AB}$. If the distance from vertex C to line of reflection $\overleftrightarrow{AB}$ is 5 cm, find the distance from vertex C' to the line of reflection, $\overleftrightarrow{AB}$.

2. Polygon CDEFG is reflected over the line of reflection, $\overleftrightarrow{AB}$. If the distance from vertex D to line of reflection $\overleftrightarrow{AB}$ is 4x cm. and the distance from vertex D' to line $\overleftrightarrow{AB}$ is 28 in, find the value of x.

3. Polygon CDEFG is reflected over the line of reflection, $\overleftrightarrow{AB}$. If the distance from vertex E to line of reflection $\overleftrightarrow{AB}$ is 3x – 20 and the distance from vertex E' to line $\overleftrightarrow{AB}$ is x + 50, find the value of x and the distance from vertex E to the line of reflection, $\overleftrightarrow{AB}$.

LAB #47: A TRANSLATION IN THE PLANE

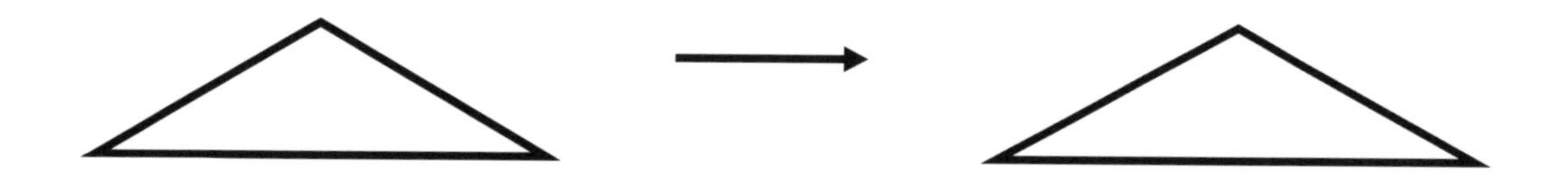

1. Select: [EDIT], [PREFERENCES], [TEXT], check the box [FOR ALL NEW POINTS] and click [OK].
2. Select: the Point tool and plot three points in the clockwise direction as the vertices of a triangle on the left part of your screen.
3. Select: the Arrow tool, [EDIT], [SELECT ALL], [CONSTRUCT], [SEGMENTS], [MEASURE], [LENGTHS] and deselect.
4. Select: the Point tool and plot a point two inches to the right of ΔABC.
5. Select: the Arrow tool and click in the blank region to deselect.
6. Select: points B, then D, [TRANSFORM] and [MARK VECTOR].
7. Select: points A, then B, then C, [TRANSFORM], [TRANSLATE], [TRANSLATE], [CONSTRUCT], [SEGMENTS], [MEASURE], [LENGTHS] and deselect.
8. Select: point D, [EDIT], [PROPERTIES], [LABEL], type a capital B' in the displayed window, click [OK] and deselect.
9. Select: points A, then B, then C, [MEASURE], [ANGLE] and deselect.
10. Do the same to measure $\angle$ ACB, $\angle$BAC, $\angle$A'B'C', $\angle$A'C'B' and $\angle$B'A'C'.
11. Click on point A of ΔABC and, keeping the left button depressed, drag it. Observe what remains equal as you do so.
12. Select: the Text tool, double click in the blank region to open a dialog box, and (1) describe your observations about the measures of sides $\overline{AB}$ and $\overline{A'B'}$, $\overline{BC}$ and $\overline{B'C'}$, $\overline{AC}$ and $\overline{A'C'}$ of ΔABC and ΔA'B'C',
    (2) describe your observations about the measures of $\angle$ABC and $\angle$A'B'C', $\angle$ACB and $\angle$A'C'B', $\angle$BAC and $\angle$B'A'C' of ΔABC and ΔA'B'C',
    (3) explain what happens to the lengths of the sides and the angle measures of the polygon under a translation.
13. Select: the Arrow tool and click in the blank region to deselect.
14. Select: points A, then B, then C, [CONSTRUCT], [TRIANGLE INTERIOR], [CONSTRUCT], [POINT ON PERIMETER], [TRANSFORM], [TRANSLATE], [TRANSLATE] and deselect.

15. Select: points E, E', [CONSTRUCT], [SEGMENT], [DISPLAY], [LINE STYLE], [THIN], [DISPLAY], [LINE STYLE], [DASHED], [MEASURE], [LENGTH] and deselect.
16. Select: point E, [EDIT], [ACTION BUTTONS], [ANIMATION], [MEDIUM], [SLOW] and [OK].
17. Click on [ANIMATE POINT] and observe the measure of the distance between points E and E' as they trace ΔABC and ΔA'B'C'.
18. Click on [ANIMATE POINT] again to stop the animation.
19. Click on point B' and, keeping the left button depressed, drag ΔA'B'C' away from ΔABC.
20. Click on [ANIMATE POINT] and observe the measure of the distance between points E and E' as they trace ΔABC and ΔA'B'C'.
21. Click on [ANIMATE POINT] again to stop the animation.
22. Select: the Text tool, double click in the blank region to open a dialog box and describe your observations about the distance between every point on ΔABC and its image on ΔA'B'C' after translation in the plane.

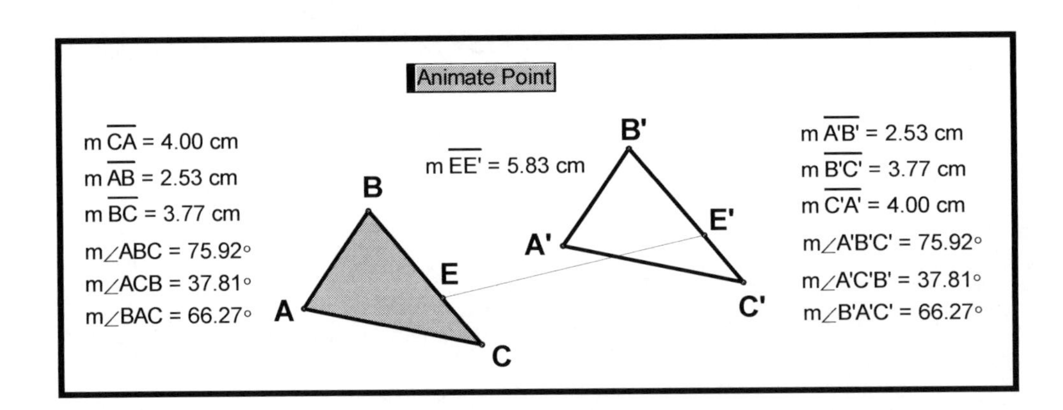

SUGGESTED EXERCISES

Type the solutions to the following problems in the blank region of your sketch.

1. ΔA'B'C' is the image of ΔABC, translated 4 inches to the left. If $m\overline{AB}$ = 1.3 in, $m\overline{BC}$ = 2.1 in, and $m\overline{AC}$ = 2.9 in respectively, find the measures of the sides $\overline{A'B'}$, $\overline{B'C'}$ and $\overline{A'C'}$ in ΔA'B'C'.
2. Find the value of x if $m\overline{CD} = 4x$ and the measure of its image $m\overline{C'D'} = 32$ cm after translation of the original quadrilateral ABCD, 5 cm to the right.
3. Find the value of x if $m\overline{CD} = 2x + 3$ and the measure of its image $m\overline{C'D'} = x + 5$ after translation 7 in. to the left of the original quadrilateral ABCD.

LAB #48: A ROTATION ABOUT A POINT

1. Select: [EDIT], [PREFERENCES], [TEXT], check the box [FOR ALL NEW POINTS] and click [OK].
2. Select: the Point tool and plot three points in the clockwise direction on the left part of your screen.
3. Select: the Arrow tool, [EDIT], [SELECT ALL], [CONSTRUCT], [SEGMENTS], [MEASURE], [LENGTHS] and deselect.
4. Select: the Point tool and plot two points one inch to the right ΔABC.
5. Select: the Arrow tool and click in the blank region to deselect.
6. Select: points A, then D, then E, [TRANSFORM] and [MARK ANGLE].
7. Select: point E, [DISPLAY] and [HIDE POINT].
8. Select: points A, then B, then C, [TRANSFORM], [ROTATE], [ROTATE], [CONSTRUCT], [SEGMENTS], [MEASURE], [LENGTHS] and deselect.
9. Select: points A, then B, then C, [MEASURE], [ANGLE] and deselect.
10. Do the same to measure $\angle ACB$ and $\angle BAC$.
11. Select: the Arrow tool, points A', then B', then C', [MEASURE], [ANGLE] and deselect.
12. Do the same to measure $\angle A'C'B'$ and $\angle B'A'C'$.
13. Click on point A of ΔABC and, keeping the left button depressed, drag it. Observe the measures that remain equal as you do so.
14. Select: the Text tool, double click in the blank region to open the dialog box, and (1) describe your observations about the measures of the sides $\overline{AB}$ and $\overline{A'B'}$, $\overline{BC}$ and $\overline{B'C'}$, $\overline{AC}$ and $\overline{A'C'}$ of ΔABC and $\Delta A'B'C'$,
    (2) describe your observations about the measures of $\angle ABC$ and $\angle A'B'C'$, $\angle ACB$ and $\angle A'C'B'$, $\angle BAC$ and $\angle B'A'C'$ of ΔABC and $\Delta A'B'C'$,
    (3) explain what happens to the lengths of the line segments and the angle measures of the polygon under a rotation.
15. Select: the Arrow tool and click in the blank region to deselect.
16. Select: points A, then B, then C, [CONSTRUCT], [TRIANGLE INTERIOR], [CONSTRUCT], [POINT ON PERIMETER], [TRANSFORM], [ROTATE], [ROTATE] and deselect.
17. Select: points D, F, [CONSTRUCT], [SEGMENT], [DISPLAY], [LINE STYLE], [THIN], [DISPLAY], [LINE STYLE], [DASHED] and deselect.

18. Select: points D, F', [CONSTRUCT], [SEGMENT] and deselect.
19. Select: points F, then D, then F', [MEASURE], [ANGLE] and deselect.
20. Select: point F, [EDIT], [ACTION BUTTONS], [ANIMATION], [MEDIUM], [SLOW] and [OK].
21. Click on [ANIMATE POINT] and observe the value of $\angle FDF'$ as points F and F' trace ΔABC and $\Delta A'B'C'$.
22. Click on [ANIMATE POINT] again to stop the animation.
23. Click on point D and keeping the left button depressed drag it to change the measure of $\angle FDF'$.
24. Click on [ANIMATE POINT] and observe the value of $\angle FDF'$ as points F and F' move around ΔABC and $\Delta A'B'C'$.
25. Click on [ANIMATE POINT] again to stop the animation.
26. Select: the Text tool, double click in the blank region to open a dialog box and describe your observations about the measure of the angle every point on ΔABC was rotated about point D to become its image on $\Delta A'B'C'$ after a rotation of the entire ΔABC about the same point in a plane

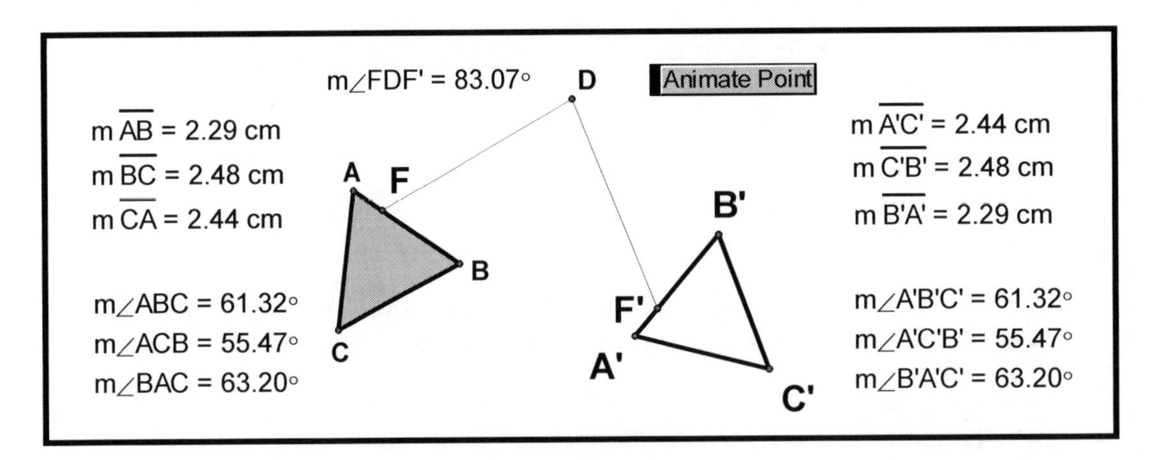

SUGGESTED EXERCISES

Type the solution to the following problems in the blank region of your sketch.

1. If the measure of $\angle A$ of ΔABC is 63°, find the measure of $\angle A'$ after a rotation of 85° about the point of rotation D.

2. Find the value of x if $m\overline{CD} = 5x$ and its image $m\overline{C'D'} = 40$ cm after rotation of the original quadrilateral ABCD 36° about point G.

3. Find the value of x if $m\overline{CD} = 3x - 3$ and its image $m\overline{C'D'} = x + 5$ after the rotation of the original polygon ABCDE 90° about the point of rotation G.

LAB #49: A DILATION ABOUT A POINT IN THE PLANE

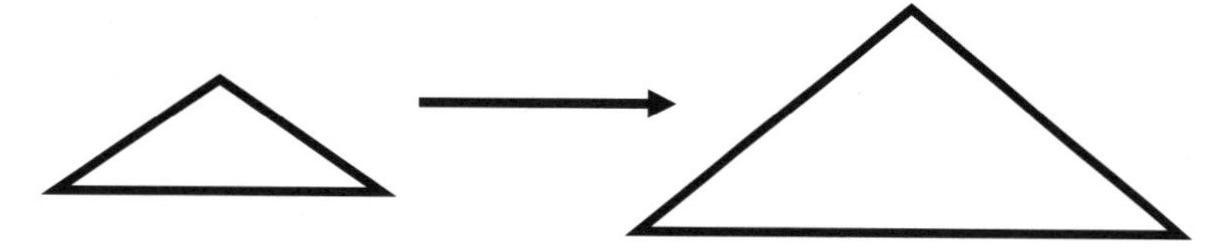

1. Select: [EDIT], [PREFERENCES], [TEXT], check the box [FOR ALL NEW POINTS] and click [OK].
2. Select: the Point tool and plot three points as the vertices of a small triangle in the middle of the left part of your screen.
3. Select: the Arrow tool, [EDIT], [SELECT ALL], [CONSTRUCT], [SEGMENTS] and deselect.
4. Select: the Point tool and plot a point approximately one inch to the left of ΔABC.
5. Select: the Arrow tool, [NUMBER], [NEW PARAMETER], type [CONSTANT OF DILATION] under [NAME], type [3] under [VALUE], [OK], [TRANSFORM] and [MARK SCALE FACTOR].
6. Select: point D, [TRANSFORM] and [MARK CENTER].
7. Select: points A, then B, then C, [TRANSFORM], [DILATE], [DILATE], [CONSTRUCT], [SEGMENTS] and deselect
8. Select: points D, then A, [CONSTRUCT], [RAY], [DISPLAY], [LINE STYLE], [THIN], [DISPLAY], [LINE STYLE], [DASHED] and deselect.
9. Do the same to construct rays $\overrightarrow{DB}$ and $\overrightarrow{DC}$.
10. Select: points A, then B, then C, [MEASURE], [ANGLE] and deselect.
11. Do the same to measure $\angle$ ACB, $\angle$BAC, $\angle$A'B'C', $\angle$A'C'B' and $\angle$B'A'C'.
12. Select: line segments $\overline{A'B'}$, then $\overline{AB}$, [MEASURE], [RATIO] and deselect.
13. Do the same to find the values of the ratios of $\overline{B'C'}$ to $\overline{BC}$ and $\overline{A'C'}$ to $\overline{AC}$.
14. Select: the Text tool, double click in the blank region to open a dialog box, and describe (1) your observations about the measures of $\angle$ABC and $\angle$A'B'C', $\angle$ACB and $\angle$A'C'B', $\angle$BAC and $\angle$B'A'C' of ΔABC and ΔA'B'C',

    (2) your observations about the values of $\frac{m\overline{A'B'}}{m\overline{AB}}$, $\frac{m\overline{C'A'}}{m\overline{CA}}$, and $\frac{m\overline{B'C'}}{m\overline{BC}}$,

    3) explain what the value of the ratio of the corresponding sides of two triangles tells you about how much every side of ΔA'B'C' is larger than the corresponding side of ΔABC.
15. Select: the Arrow tool and click in the blank region to deselect.
16. Click on point C', and keeping the left button depressed, drag it. Observe the values that remain constant as you do so.

17. Select: the Text tool, double click in the blank region to open a dialog box and (1) describe your observations of the value of the constant of dilation and the values of the ratios of the corresponding sides of ΔABC and $\Delta A'B'C'$, (2) explain what you have to do with the corresponding sides of ΔABC and $\Delta A'B'C'$ to find the constant of dilation.
18. Select: the caption that shows [CONSTANT OF DILATION = 3], [EDIT], [ACTION BUTTONS], [ANIMATION], type [2] in the window marked as [SEC], type [0.3] in the left window marked as [DOMAIN], type [3.7] in the right window marked as [DOMAIN], click [OK] and deselect.
19. Click on [ANIMATE PARAMETER] and observe the values that remain equal.
20. Click on [ANIMATE PARAMETER] again to stop the animation.
21. Select: the Text tool, double click in the blank region to open a dialog box and (1) explain what happens to the angle measures of ΔABC under a dilation about point D in a plane, (2) describe your observation about the constant of dilation and the ratios of the corresponding sides of ΔABC and $\Delta A'B'C'$, (3) choose the appropriate word to describe ΔABC and its dilated image $\Delta A'B'C'$: (a) congruent (b) similar.

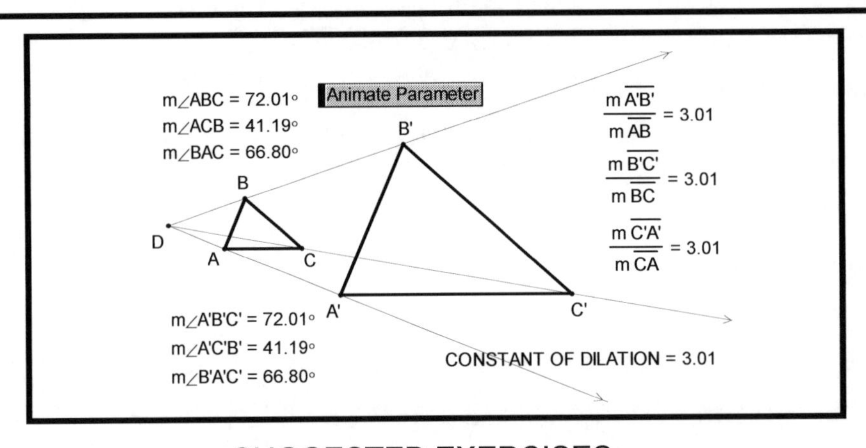

SUGGESTED EXERCISES

Type the solution to the following problems in the blank region of your sketch.

1. Find the measures of the images of the sides of ΔABC if $m\overline{AB} = 5$ cm, $m\overline{BC} = 7$ cm, and $m\overline{AC} = 11$ cm of after a dilation in a plane if the constant of dilation is 4.

2. Side $\overline{A'B'}$ of $\Delta A'B'C'$ is the dilated image of side $\overline{AB}$ of ΔABC. If $m\overline{A'B'} = 12$ and the constant of dilation is 2, find the measure of the side $\overline{AB}$ before the dilation.

3. Side $\overline{A'B'}$ of $\Delta A'B'C'$ is the dilated image of side $\overline{AB}$ of ΔABC. If $m\overline{A'B'} = 15$ cm, and $m\overline{AB} = 3$ cm, find the constant of dilation.

4. Side $\overline{A'B'}$ of $\Delta A'B'C'$ is the dilated image of side $\overline{AB}$ of ΔABC. If $m\overline{A'B'} = 3$ cm, and $m\overline{AB} = 9$ cm, find the constant of dilation.

LAB #50: THE MEDSEGMENT OF A TRAPEZOID

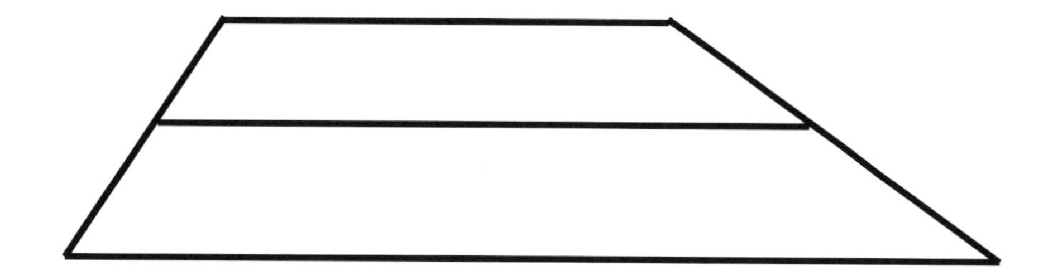

1. Select: [EDIT], [PREFERENCES], [TEXT], check the box [FOR ALL NEW POINTS] and click [OK].
2. Select: the Line Segment tool and construct a horizontal line segment.
3. Select: the Point tool and plot a point anywhere above line segment $\overline{AB}$.
4. Select: the Arrow tool, [EDIT], [SELECT ALL], [EDIT], [SELECT CHILDREN], [CONSTRUCT], [PARALLEL LINE], [CONSTRUCT] and [POINT ON PARALLEL LINE].
5. Click on point D, and keeping the left button depressed, drag it one inch to the left of point C and deselect.
6. Select: line $\overleftrightarrow{CD}$, [DISPLAY] and [HIDE PARALLEL LINE].
7. Select: points A, then B, then C, then D, [CONSTRUCT], [SEGMENTS], [MEASURE], [LENGTHS] and deselect.
8. Select: line segment $\overline{AD}$, line segment $\overline{BC}$, [CONSTRUCT], [MIDPOINTS], [CONSTRUCT], [SEGMENT], [MEASURE], [LENGTH] and deselect.
9. Select: [NUMBER], [CALCULATE], [( ], click on the caption that shows $m\overline{CD}$, [+], click on the caption that shows $m\overline{AB}$, [ )], [÷], [2], [OK] and deselect. (The average measure of bases $\overline{CD}$ and $\overline{AB}$ will appear on your sketch. Compare this measure to the measure of median $\overline{EF}$.)
10. Click on any labeled point, and keeping the left button depressed, drag it. Observe the values that remain equal as you do so.

11. Select: the Text tool, double click in the blank region to open a dialog box and explain how to find the measure of the midsegment of the trapezoid using its upper and the lower bases.

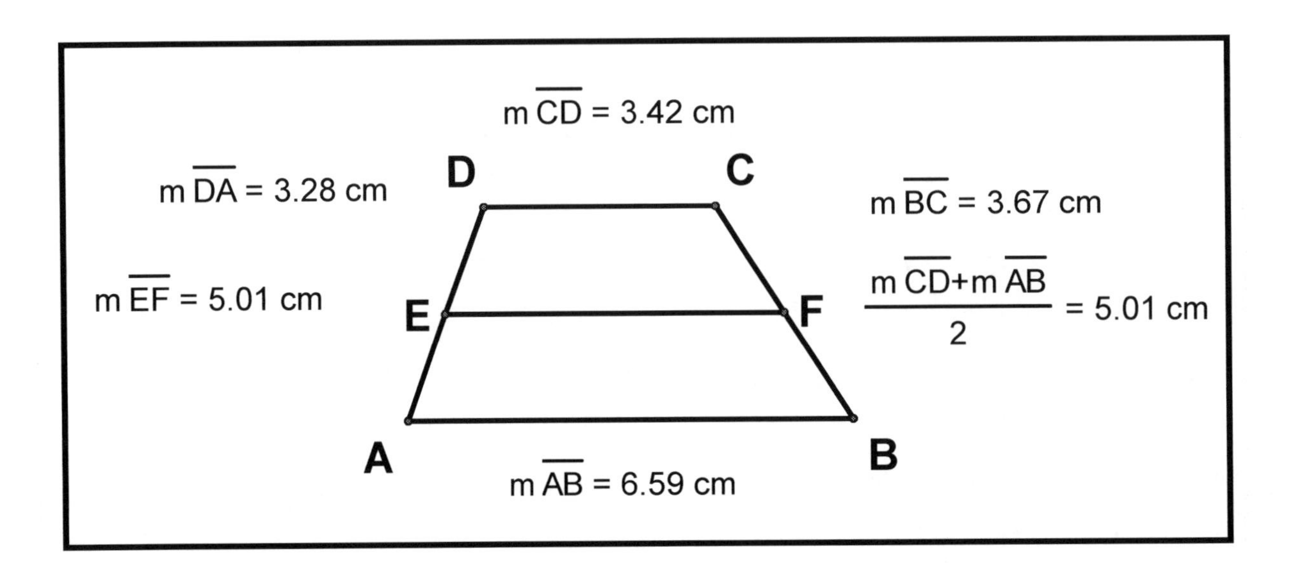

SUGGESTED EXERCISES

Type the solutions to the following problems in the blank region of your sketch.

1. $\overline{AB}$ and $\overline{CD}$ are the bases of trapezoid ABCD. If $m\overline{CD}$ = 12 in, and $m\overline{AB}$ = 20 in, find the measure of midsegment $\overline{EF}$.

2. $\overline{EF}$ is the midsegment of trapezoid ABCD. If $m\overline{EF}$ = 8 cm and $m\overline{CD}$ = 5 cm, find the measure of the base $\overline{AB}$.

3. $\overline{EF}$ is the midsegment of trapezoid ABCD. If $m\overline{EF}$ = 7 cm, $m\overline{CD}$ = 2x + 1 and $m\overline{AB}$ = 3x – 17, find the value of x and the measure of each base.

LAB #51: THE AREA OF A SQUARE

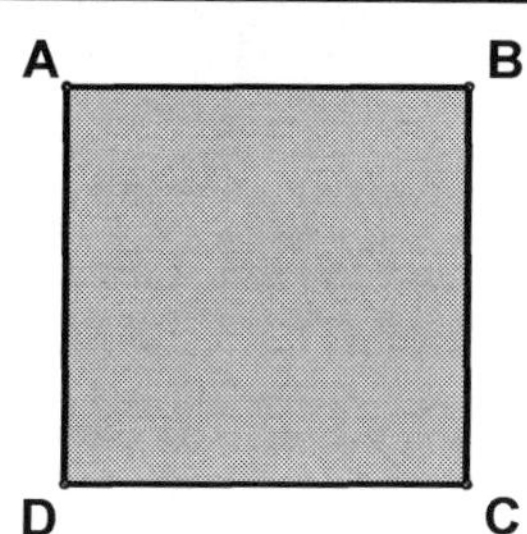

1. Select: [EDIT], [PREFERENCES], [TEXT], check the box [FOR ALL NEW POINTS] and click [OK].
2. Select: the Point tool and plot two points approximately one inch apart from each other in the middle of your screen.
3. Select: the Arrow tool and click in the blank region to deselect.
4. Select: point B, [TRANSFORM] and [MARK CENTER].
5. Select: point A, [TRANSFORM], [ROTATE], [ROTATE] and deselect.
6. Select: point A', [EDIT], [PROPERTIES], [LABEL], type a capital C in the displayed window, click [OK] and deselect.
7. Select: point C, [TRANSFORM] and [MARK CENTER].
8. Select: point B, [TRANSFORM], [ROTATE], [ROTATE] and deselect.
9. Select: point B', [EDIT], [PROPERTIES], type a capital D in the displayed window, click [OK] and deselect.
10. Select: [EDIT], [SELECT ALL], [CONSTRUCT], [SEGMENTS], [MEASURE] and [LENGTHS].
11. Select: the Text tool, double click in the blank region to open a dialog box and describe your observations about the measures of the sides of the square.
12. Select: the Arrow tool and click in the blank region to deselect.
13. Select: points A, then B, then C, then D, [CONSTRUCT], [QUADRILATERAL INTERIOR], [MEASURE], [AREA] and deselect.

14. Select: [NUMBER], [CALCULATE], click on the caption that shows the measure of any side of the square, [∧], [2], [OK] and deselect.
15. Click on point B and, keeping the left button depressed, drag it. Observe the values that remain equal as you do so.
16. Select: the Text tool, double click in the blank region to open a dialog box and explain how to find the area of a square using any of its sides.

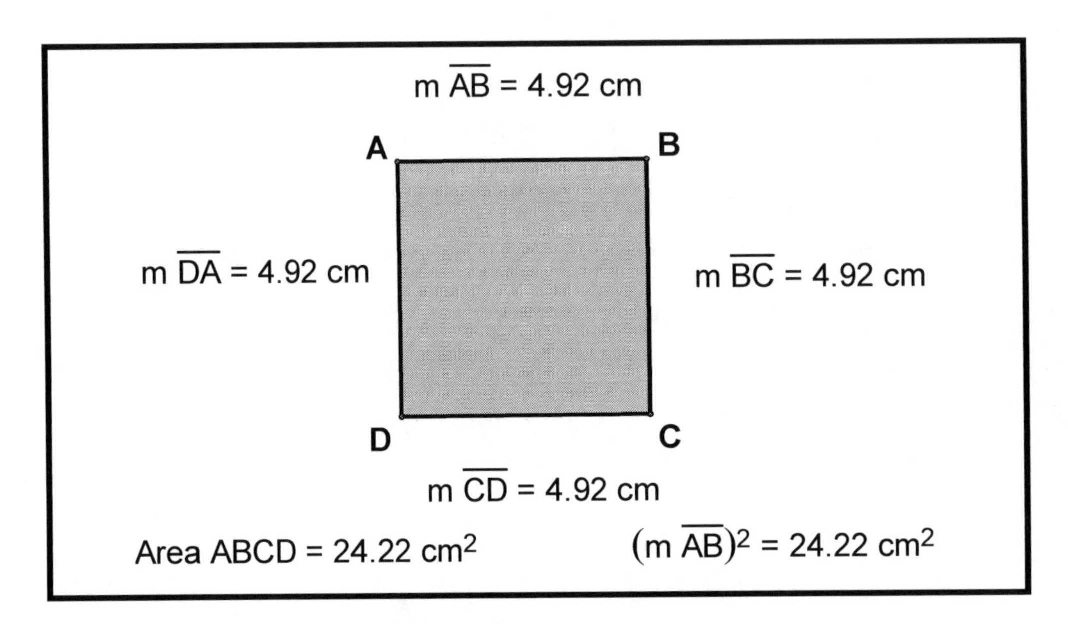

SUGGESTED EXERCISES

Type the solutions to the following problems in the blank region of your sketch.

1. Find the area of a square whose side is 7 in.
2. Express in terms of x the area of a square whose side is 2x.
3. Express in terms of x the area of a square whose side is (x + 3).
4. Find the measure of the side of a square whose area is 36 square inches.
5. Express in terms of x the side of a square whose area is $16x^2$.
6. Express in terms of x the side of a square whose area is $x^2 + 10x + 25$.

LAB #52: THE AREA OF A RECTANGLE

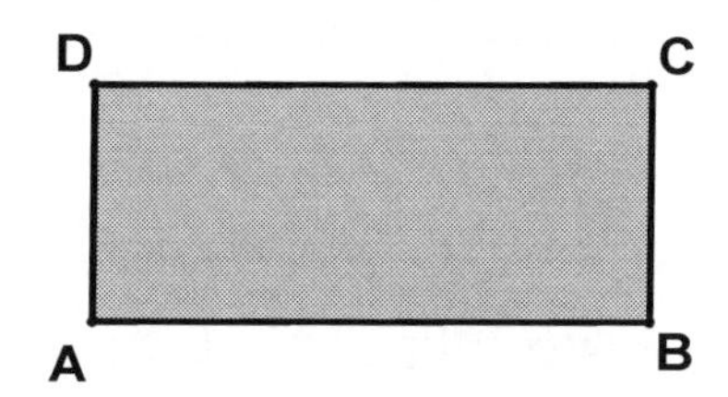

1. Select: [EDIT], [PREFERENCES], [TEXT], check the box [FOR ALL NEW POINTS] and click [OK].
2. Select: the Line Segment tool and construct a horizontal line segment in the middle of your screen.
3. Select: the Arrow tool, [EDIT], [SELECT ALL], [CONSTRUCT], [PERPENDICULAR LINES] and deselect.
4. Select: the perpendicular line passing through point B, [CONSTRUCT] and [POINT ON PERPENDICULAR LINE].
5. Click on point C, and keeping the left button depressed, drag it one inch above point B and deselect.
6. Select: point C, line $\overleftrightarrow{BC}$, [CONSTRUCT], [PERPENDICULAR LINE] and deselect.
7. Select: the perpendicular line passing through point A, the perpendicular line passing through point C, [CONSTRUCT], [INTERSECTION] and deselect.
8. Select: the Arrow tool, lines $\overleftrightarrow{AD}$, $\overleftrightarrow{DC}$, $\overleftrightarrow{BC}$, [DISPLAY] and [HIDE PERPENDICULAR LINES].
9. Select: points A, then D, then C, then B, [CONSTRUCT], [SEGMENTS], [MEASURE], [LENGTHS] and deselect.
10. Select: points A, then B, then C, then D, [CONSTRUCT], [QUADRILATERAL INTERIOR], [MEASURE], [AREA] and deselect.
11. Select: [NUMBER], [CALCULATE], click on the caption that shows $m\overline{AB}$, [*], click on the caption that shows $m\overline{AD}$, [OK] and deselect.

12. Do the same to multiply base $\overline{DC}$ and altitude $\overline{CB}$.

13. Select: the Arrow tool, click on point A, B or C, and keeping the left button depressed, drag it. Observe the values that remain equal as you do so.

14. Select: the Text tool, double click in the blank region to open a dialog box and explain how to find the area of a rectangle using its base and altitude.

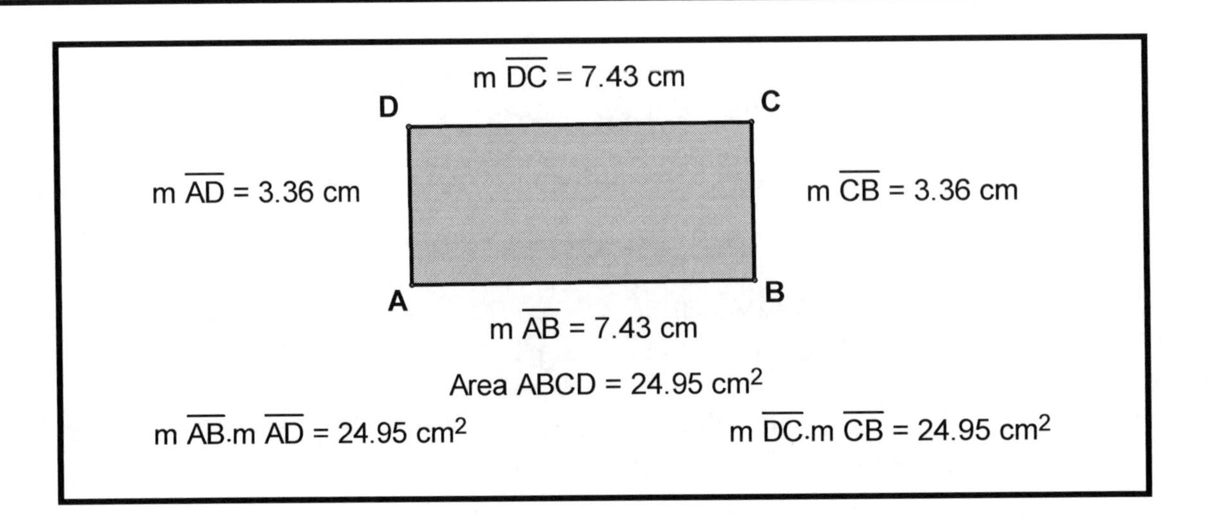

SUGGESTED EXERCISES

Type the solutions to the following problems in the blank region of your sketch.

1. Find the area of the rectangle ABCD if its altitude is 6 ft and its base is 13 ft.

2. Find the base of the rectangle ABCD if its area is 87 ft^2 and its altitude is 6 ft.

3. The area of rectangle ABCD is 63 cm^2. If the measure of its base $\overline{AB}$ is 9 cm and the measure of its altitude $\overline{BC}$ is $2x - 3$, find the value of x and the measure of the altitude.

4. The area of rectangle ABCD is 98 cm^2. If the measure of its base $\overline{AB}$ is 2x and the measure of its altitude $\overline{AD}$ is x, find the measure of its altitude and the measure of its base.

LAB #53: THE AREA OF A PARALLELOGRAM

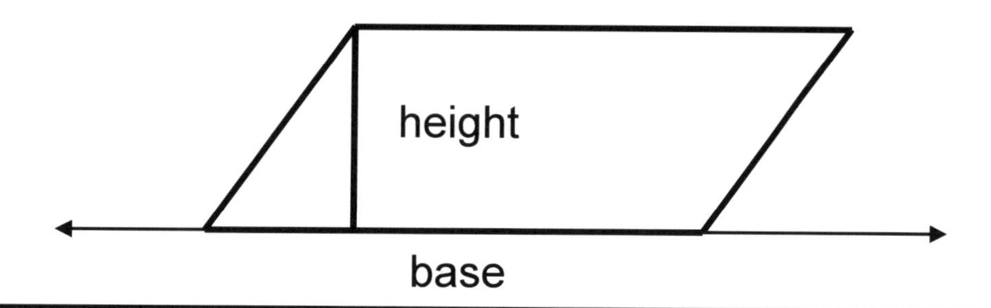

1. Select: [EDIT], [PREFERENCES], [TEXT], check the box [FOR ALL NEW POINTS] and click [OK].
2. Select: the Point tool and plot two points as the endpoints of a horizontal line segment in the middle of your screen.
3. Select: the Arrow tool, [EDIT], [SELECT ALL], [CONSTRUCT], [LINE], [DISPLAY], [LINE STYLE], [THIN], [DISPLAY], [LINE STYLE], [DASHED] and deselect.
4. Select: the Point tool and plot a point anywhere above line $\overleftrightarrow{AB}$.
5. Select: the Arrow tool and click in the blank region to deselect.
6. Select: points B, C, [CONSTRUCT], [SEGMENT] and deselect.
7. Select: point C, line $\overleftrightarrow{AB}$, [CONSTRUCT], [PARALLEL LINE] and deselect.
8. Select: point A, line segment $\overline{BC}$, [CONSTRUCT], [PARALLEL LINE] and deselect.
9. Select: the parallel line passing through point A, the parallel line passing through point C, [CONSTRUCT], [INTERSECTION] and deselect.
10. Select: point C, line $\overleftrightarrow{AB}$, [CONSTRUCT], [PERPENDICULAR LINE], line $\overleftrightarrow{AB}$, [CONSTRUCT], [INTERSECTION] and deselect.
11. Select: lines $\overleftrightarrow{CD}$, $\overleftrightarrow{AD}$, $\overleftrightarrow{CE}$ [DISPLAY] and [HIDE LINES].
12. Select: points A, then D, then C, then B, [CONSTRUCT], [SEGMENTS], [DISPLAY], [LINE STYLE], [MEDIUM], [DISPLAY], [LINE STYLE], [SOLID], [MEASURE], [LENGTHS] and deselect.
13. Select points C, E, [CONSTRUCT], [SEGMENT], MEASURE], [LENGTH] and deselect.

14. Select: points A, then B, then C, then D, [CONSTRUCT], [QUADRILATERAL INTERIOR], [MEASURE], [AREA] and deselect.
15. Select: [NUMBER], [CALCULATE], click on the caption that shows $m\overline{BA}$, [∗], click on the caption that shows $m\overline{CE}$, [OK] and deselect.
16. Do the same to multiply the upper base $\overline{DC}$ and altitude $\overline{CE}$.
17. Click on point C, and keeping the button depressed, drag it. Then click on point B and do the same. Observe the values that remain equal as you do so.
18. Select: the Text tool, double click in the blank region to open a dialog box, and explain how to find the area of a parallelogram using its base and altitude.

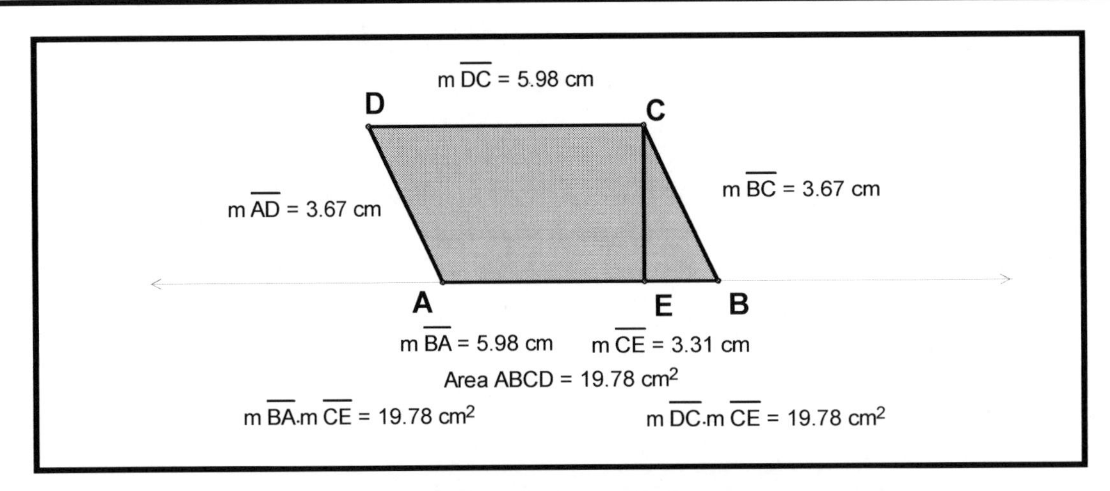

SUGGESTED EXERCISES

Type the solutions to the following problems in the blank region of your sketch.

1. Find the area of parallelogram ABCD if the measure of its altitude is 7 ft and the measure of its base is 15 ft.
2. The area of parallelogram ABCD is 60 in^2. If the measure of its base $\overline{AB}$ is 12 in and the measure of its height $\overline{CE}$ is x, find the height of the parallelogram.
3. The area of parallelogram ABCD is 70 in^2. If the measure of its height $\overline{CE}$ is 7 in and the measure of its base $\overline{AB}$ is $x - 2$, find the value of x and the measure of the base.

LAB #54: THE AREA OF A TRIANGLE

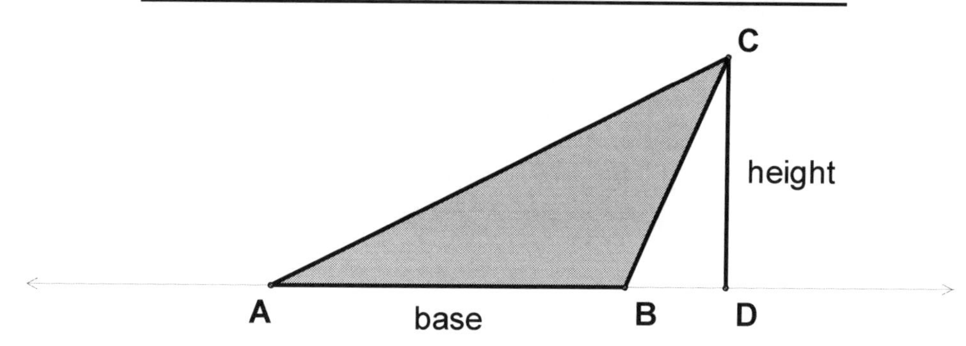

1. Select: [EDIT], [PREFERENCES], [TEXT], check the box [FOR ALL NEW POINTS] and click [OK].
2. Select: the Point tool and plot two points as the endpoints of a horizontal line segment.
3. Select: the Arrow tool, [EDIT], [SELECT ALL], [CONSTRUCT], [LINE], [DISPLAY], [LINE STYLE], [THIN], [DISPLAY], [LINE STYLE] and [DASHED].
4. Select: the Point tool and plot a point anywhere above line $\overleftrightarrow{AB}$.
5. Select: the Arrow tool and click in the blank region to deselect.
6. Select: point C, line $\overleftrightarrow{AB}$, [CONSTRUCT], [PERPENDICULAR LINE] and deselect.
7. Select: the perpendicular line passing through point C, line $\overleftrightarrow{AB}$, [CONSTRUCT], [INTERSECTION] and deselect.
8. Select: line $\overleftrightarrow{CD}$, [DISPLAY] and [HIDE PERPENDICULAR LINE].
9. Select: points A, then B, then C, [CONSTRUCT], [SEGMENTS], [DISPLAY], [LINE STYLE], [SOLID], [DISPLAY], [LINE STYLE], [MEDIUM], [MEASURE], [LENGTHS] and deselect.
10. Select: points C, D, [CONSTRUCT], [SEGMENT], [MEASURE], [LENGTH] and deselect.
11. Select: points A, then B, then C, [CONSTRUCT], [TRIANGLE INTERIOR], [MEASURE], [AREA] and deselect.
12. Select: [NUMBER], [CALCULATE], click on the caption that shows $m\overline{AB}$, [$*$], click on the caption that shows $m\overline{CD}$, [$\div$], [2], [OK] and deselect.

13. Select: the Arrow tool, click on point C and, keeping the left button depressed, drag it. Then click on point B and do the same. Observe the values that remain equal as you do so.
14. Select: the Text tool, double click in the blank region to open a dialog box and explain how to find the area of the triangle using its base and altitude.

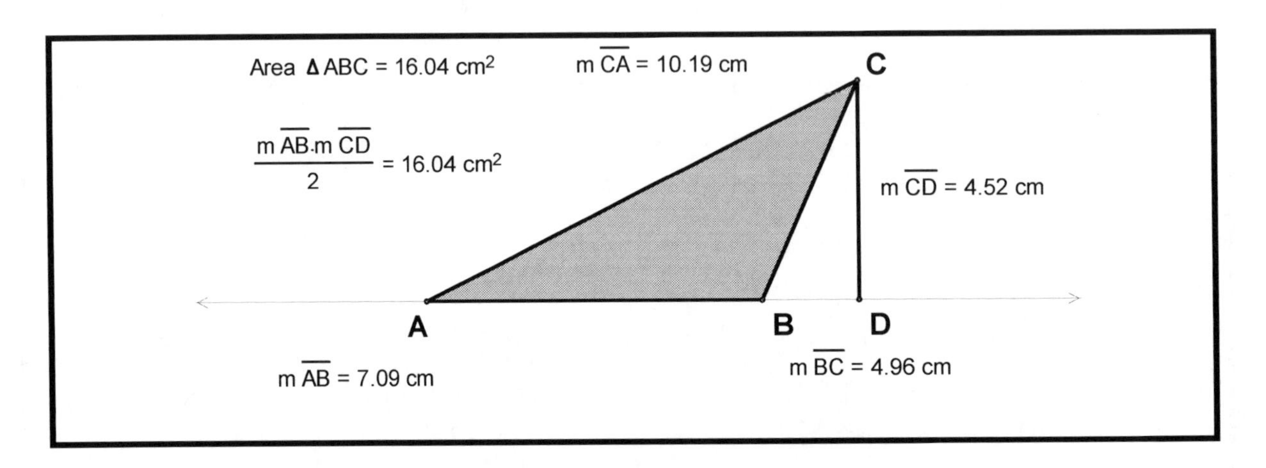

SUGGESTED EXERCISES

Type the solutions to the following problems in the blank region of your sketch.

1. Find the area of the ΔABC if its altitude is equal 5 ft and its base is equal to 12 ft.

2. Find the length of altitude $\overline{CD}$ of ΔABC whose area 40 ft^2 and whose base is 10 ft.

3. The area of ΔABC is 36 cm^2. If the measure of altitude $\overline{CD}$ is 2x and the measure of base $\overline{AB}$ is 4x, find the value of x and the measures of altitude $\overline{CD}$ and of base $\overline{AB}$.

4. The area of ΔABC is 24 cm^2. If the measure of altitude $\overline{CD}$ is 2x and the measure of base $\overline{AB}$ is $x - 2$, find the value of x and the measures of altitude $\overline{CD}$ and of base $\overline{AB}$.

LAB #55: THE AREA OF A TRAPEZOID

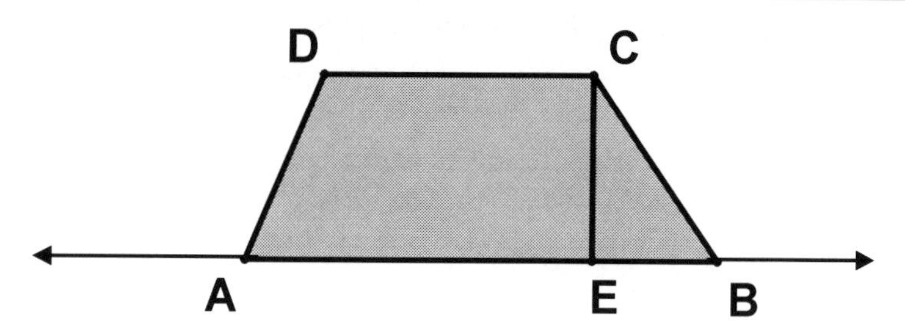

1. Select: [EDIT], [PREFERENCES], [TEXT], check the box [FOR ALL NEW POINTS] and click [OK].
2. Select: the Point tool and plot two points as the endpoints of a horizontal line segment.
3. Select: the Arrow tool, [EDIT], [SELECT ALL], [CONSTRUCT], [LINE], [DISPLAY], [LINE STYLE], [THIN], [DISPLAY], [LINE STYLE], [DASHED] and deselect.
4. Select: the Point tool and plot a point anywhere above line $\overleftrightarrow{AB}$.
5. Select: the Arrow tool and click in the blank region to deselect.
6. Select: point C, line $\overleftrightarrow{AB}$, [CONSTRUCT], [PARALLEL LINE], [CONSTRUCT] and [POINT ON PARALLEL LINE].
7. Click on point D and, keeping the left button depressed, drag it one inch to the left of point C and deselect.
8. Select: point C, line $\overleftrightarrow{AB}$, [CONSTRUCT], [PERPENDICULAR LINE], line $\overleftrightarrow{AB}$, [CONSTRUCT], [INTERSECTION] and deselect.
9. Select: lines $\overleftrightarrow{CD}$, $\overleftrightarrow{CE}$, [DISPLAY] and [HIDE LINES].
10. Select: points A, then D, then C, then B, [CONSTRUCT], [SEGMENTS], [DISPLAY], [LINE STYLE], [SOLID], [DISPLAY], [LINE STYLE], [DASHED], [MEASURE], [LENGTHS] and deselect.
11. Select: points C, E, [CONSTRUCT], [SEGMENT], [MEASURE], [LENGTH] and deselect.
12. Select: points A, then B, then C, then D, [CONSTRUCT], [QUADRILATERAL INTERIOR], [MEASURE], [AREA] and deselect.

13. Select: [NUMBER], [CALCULATE], the caption that shows $m\overline{CE}$, [*], [( ], the caption that shows $m\overline{BA}$, [+], the caption that shows $m\overline{DC}$, [ )], [÷], [2], [OK] and deselect.
14. Click on point C and, keeping the left button depressed, drag it. Then click on point B and do the same. Observe the values that remain equal as you do so.
15. Select: the Text tool, double click in the blank region to open a dialog box and explain how to find the area of the trapezoid using its altitude and both of its bases.

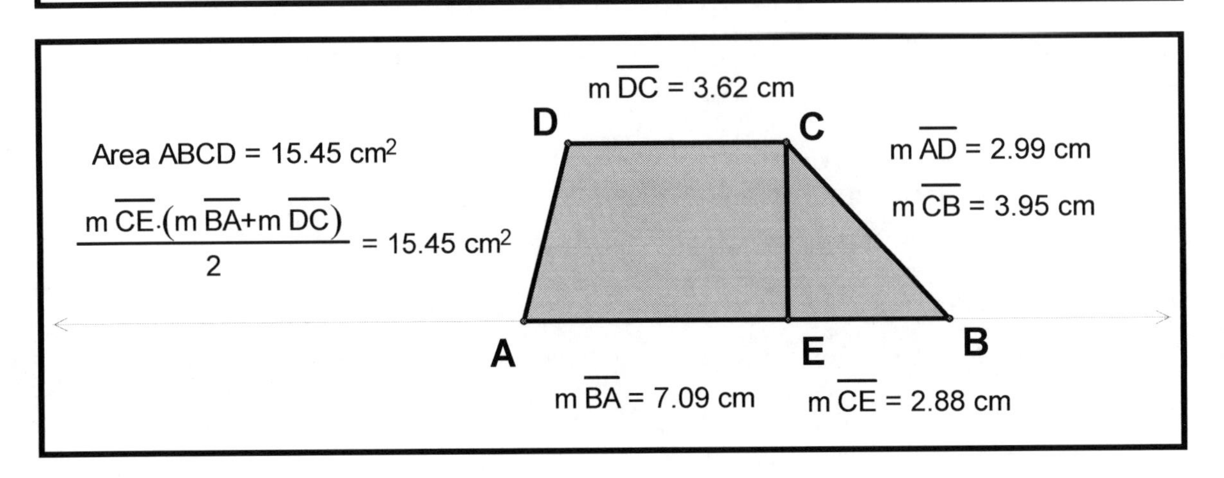

SUGGESTED EXERCISES

Type the solutions to the following problems in the blank region of your sketch.

1. Find the area of trapezoid ABCD if its altitude is 7 ft and its bases are 12 ft and 16 ft, respectively.

2. The area of trapezoid ABCD is 46 in^2. If the measure of altitude $\overline{CE}$ is 4 in, the measure of base $\overline{AB}$ is 6 in, and the measure of base $\overline{DC}$ is x, find the measure of base $\overline{CD}$.

3. The area of trapezoid ABCD is 48 sq in. If the measure of altitude $\overline{CE}$ is 6 in, the measure of base $\overline{AB}$ is x + 4 and the measure of base $\overline{DC}$ is 5x – 6, find the value of x and the measure of each base.

LAB #56: THE AREA OF A SQUARE USING ITS DIAGONALS

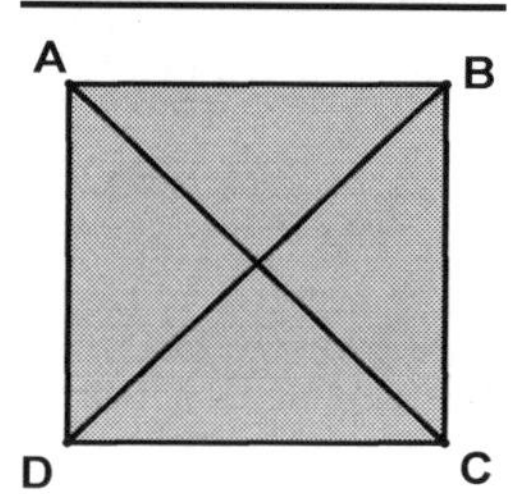

1. Select: [EDIT] [PREFERENCES], [TEXT], check the box [FOR ALL NEW POINTS] and click [OK].
2. Select: the Point tool and plot two points approximately one inch apart from each other in the middle of your screen as the endpoints of a horizontal line segment.
3. Select: the Arrow tool and click in the blank region to deselect.
4. Select: point B, [TRANSFORM] and [MARK CENTER].
5. Select: point A, [TRANSFORM], [ROTATE], [ROTATE] and deselect.
6. Select: point A', [EDIT], [PROPERTIES], [LABEL], type a capital C in the displayed window, click [OK] and deselect.
7. Select: point C, [TRANSFORM] and [MARK CENTER].
8. Select: point B, [TRANSFORM], [ROTATE], [ROTATE] and deselect.
9. Select: point B', [EDIT], [PROPERTIES], type a capital D in the displayed window, click [OK] and deselect.
10. Select: [EDIT], [SELECT ALL], [CONSTRUCT], [SEGMENTS] and deselect.
11. Select: points A, then C, then B, then D, [CONSTRUCT], [SEGMENTS] and deselect.
12. Select: diagonals $\overline{AC}$, $\overline{BD}$, [MEASURE], [LENGTH] and deselect.
13. Select: points A, then B, then C, then D, [CONSTRUCT], [QUADRILATERAL INTERIOR], [MEASURE], [AREA] and deselect.

14. Select: [NUMBER], [CALCULATE], click on the caption that shows m$\overline{AC}$, [∧], [2], [÷], [2], [OK] and deselect.
15. Select: [NUMBER], [CALCULATE], click on the caption that shows m$\overline{BD}$, [∧], [2], [÷], [2] and [OK].
16. Select: the Text tool, double click in the blank region to open a dialog box and describe your observations about the measure of the area of the square ABCD and the values of $\frac{(m\overline{AC})^2}{2}$ and $\frac{(m\overline{BD})^2}{2}$.
17. Select: the Arrow tool, click on point A, B or C, and keeping the left button depressed, drag it. Observe the values that remain equal as you do it.
18. Select: the Text tool, double click in the blank region to open a dialog box and explain how to find the area of a square using its diagonals.

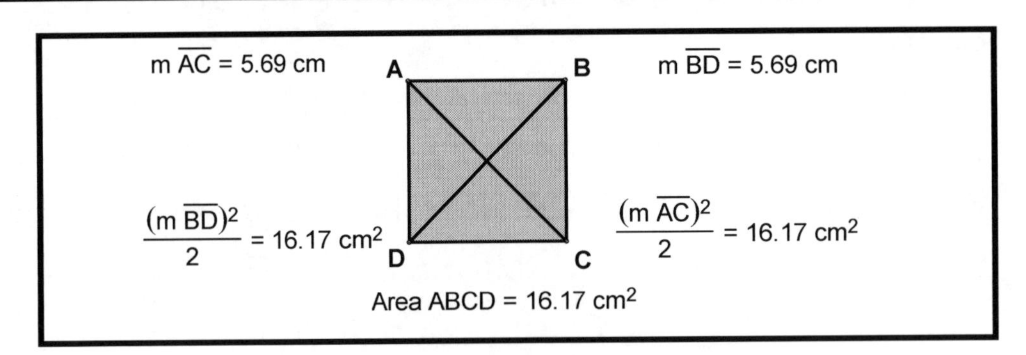

SUGGESTED PROBLEMS

Type the solutions to the following problems in the blank region of your sketch.

1. Find the area of the square ABCD if its diagonals are equal to 7 ft.
2. The area of the square ABCD is 72 in^2. If the measure of diagonal $\overline{AC}$ is x, and the measure of diagonal $\overline{BD}$ is x, find the measure of each of the diagonals.
3. The area of the square ABCD is 32 in^2. If the measure of diagonal $\overline{AC}$ is 2x, and the measure of diagonal $\overline{BD}$ is 2x, find the measure of each of the diagonals.

LAB #57: DISTINGUISHING BETWEEEN THE AREA AND PERIMETER OF A TRIANGLE

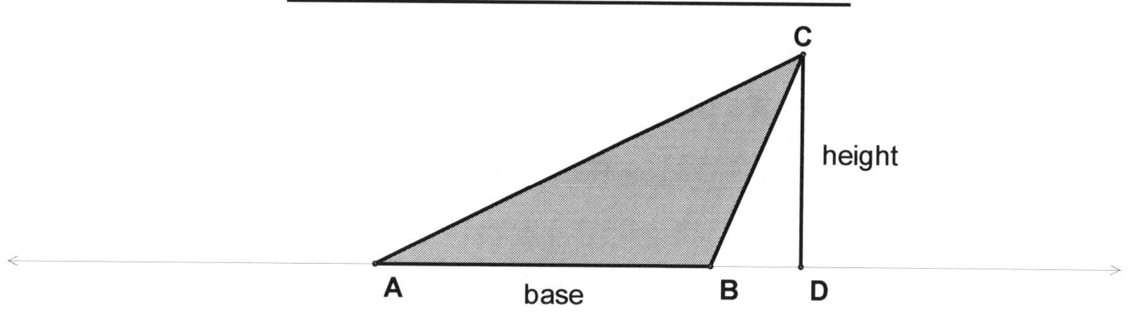

1. Select: [EDIT], [PREFERENCES], [TEXT], check the box [FOR ALL NEW POINTS] and click [OK].
2. Select: the Point tool and plot two points as the endpoints of a horizontal line segment.
3. Select: the Arrow tool, [EDIT], [SELECT ALL], [CONSTRUCT], [LINE], [DISPLAY], [LINE STYLE], [THIN], [DISPLAY], [LINE STYLE] and [DASHED].
4. Select: the Point tool and plot a point anywhere above line $\overleftrightarrow{AB}$.
5. Select: the Arrow tool and click in the blank region to deselect.
6. Select: point C, line $\overleftrightarrow{AB}$, [CONSTRUCT], [PERPENDICULAR LINE], line $\overleftrightarrow{AB}$, [CONSTRUCT], [INTERSECTION] and deselect.
7. Select: line $\overleftrightarrow{CD}$, [DISPLAY] and [HIDE PERPENDICULAR LINE].
8. Select: points A, then B, then C, [CONSTRUCT], [SEGMENTS], [DISPLAY], [LINE STYLE], [SOLID], [DISPLAY], [LINE STYLE], [DASHED], [MEASURE], [LENGTHS] and deselect.
9. Select: points C, D, [CONSTRUCT], [SEGMENT], [MEASURE], [LENGTH] and deselect.
10. Select: points A, then B, then C, [CONSTRUCT], [TRIANGLE INTERIOR], [MEASURE], [AREA], [EDIT], [SELECT PARENTS], [MEASURE], [PERIMETER] and deselect.
11. Select: the Text tool, double click in the blank region to open a dialog box and (1) explain why the area and the perimeter of ΔABC are two different numbers, (2) explain the difference between the units of measurement that represent the area and the perimeter.

12. Select: [NUMBER], [CALCULATE], click on the caption that shows $m\overline{AB}$, [*], click on the caption that shows $m\overline{CD}$, [÷], [2], [OK] and deselect.
13. Select: [NUMBER], [CALCULATE], click on the caption that shows $m\overline{AB}$, [+], click on the caption that shows $m\overline{BC}$, [+], click on the caption that shows $m\overline{CA}$, [OK] and deselect.
14. Select: the Arrow tool, click on point C, and keeping the left button depressed, drag it. Then click on point B and do the same. Observe the values that remain equal as you do so.
15. Select: the Text tool, double click in the blank region to open a dialog box and explain (1) how to find the area of the triangle using its base and height, (2) how to find the perimeter of the triangle using its sides, (3) the difference between the area and the perimeter of a triangle.

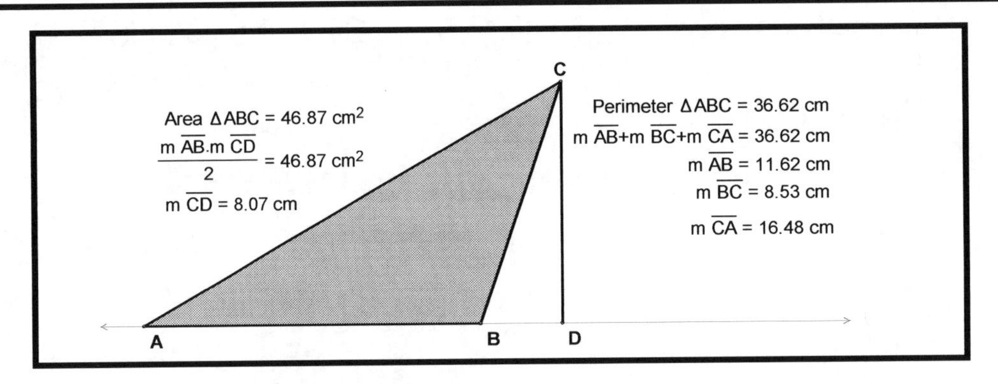

SUGGESTED EXERCISES

Type the solutions to the following problems in the blank region of your sketch.

1. In equilateral ΔABC, $m\overline{AB} = 10$ in, $m\overline{BC} = 10$ in, $m\overline{AC} = 10$ in. The altitude $\overline{CD}$ is drawn from the vertex C to the base $\overline{AB}$. If $m\overline{CD} = 5\sqrt{3}$ in, find (a) the area of ΔABC and (b) the perimeter of ΔABC.

2. In right ΔABC the measure of leg $\overline{AB}$ is 3 in and the measure of leg $\overline{AC}$ is 4 in. (a) Find the hypotenuse of the triangle, (b) find the perimeter of the triangle, and (c) find the area of the triangle. (Which side is the base and which side is the altitude?)

3. The perimeter of isosceles ΔABC is 50 cm. If $m\overline{BC} = 17$ cm, $m\overline{AC} = 17$ cm, and the measure of altitude $\overline{CD} = 15$ cm, find (a) the base AB, and (b) the area of ΔABC.

LAB #58: THE RADIUS AND THE DIAMETER OF A CIRCLE

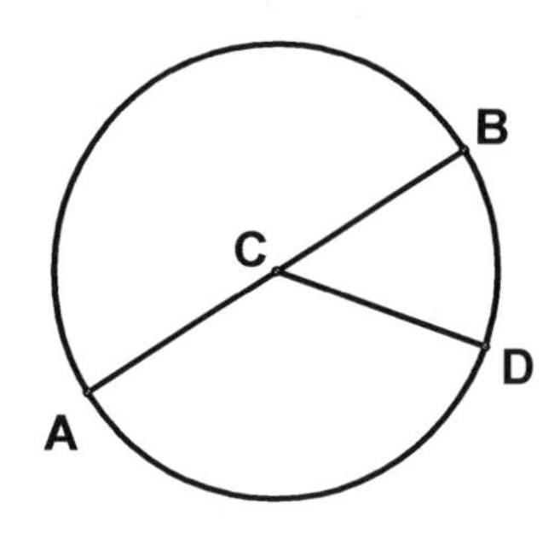

1. Select: [EDIT], [PREFERENCES], [TEXT], check the box [FOR ALL NEW POINTS] and click [OK].
2. Select: the Point tool and plot two points about two inches apart from each other in the middle of your screen.
3. Select: the Arrow tool, [EDIT], [SELECT ALL], [CONSTRUCT], [SEGMENT], [MEASURE], [LENGTH], [EDIT], [SELECT PARENTS], [CONSTRUCT], [MIDPOINT] and deselect
4. Select: points C, then B, [CONSTRUCT], [CIRCLE BY CENTER + POINT], [MEASURE], [RADIUS], [EDIT], [SELECT PARENTS], [CONSTRUCT], [POINT ON CIRCLE] and deselect.
5. Select: points C, then D, [CONSTRUCT], [SEGMENT], [MEASURE] and [LENGTH].
6. Select: the Text tool, double click in the blank region to open a dialog box and describe your observations about the measure of line segment $\overline{CD}$ and the measure of the radius of the circle.
7. Select: the Arrow tool, click on point D and, keeping the left button depressed, drag it. Observe the values that remain equal as you do so.
8. Select: the Text tool, double click in the blank region to open a dialog box and explain what two points a line segment must connect to be called a radius of a circle.
9. Select: [NUMBER], [CALCULATE], [2], [*], click on the caption that shows $m\overline{CD}$ and [OK].

10. Select: the Text tool, double click in the blank region to open a dialog box and describe your observations about how the measure of $2 \cdot m\overline{CD}$ compares to the measure of the diameter $\overline{AB}$.

11. Select: the Arrow tool, click on point B and, keeping the button depressed, drag it. Observe what values remain equal as you do so.

12. Select: the Text tool, double click in the blank region to open a dialog box and (1) explain how to find the measure of the diameter of the circle using its radius, (2) given that points A and B are two points on the circle, explain through what point the line segment $\overline{AB}$ must pass in order to be called a diameter of a circle.

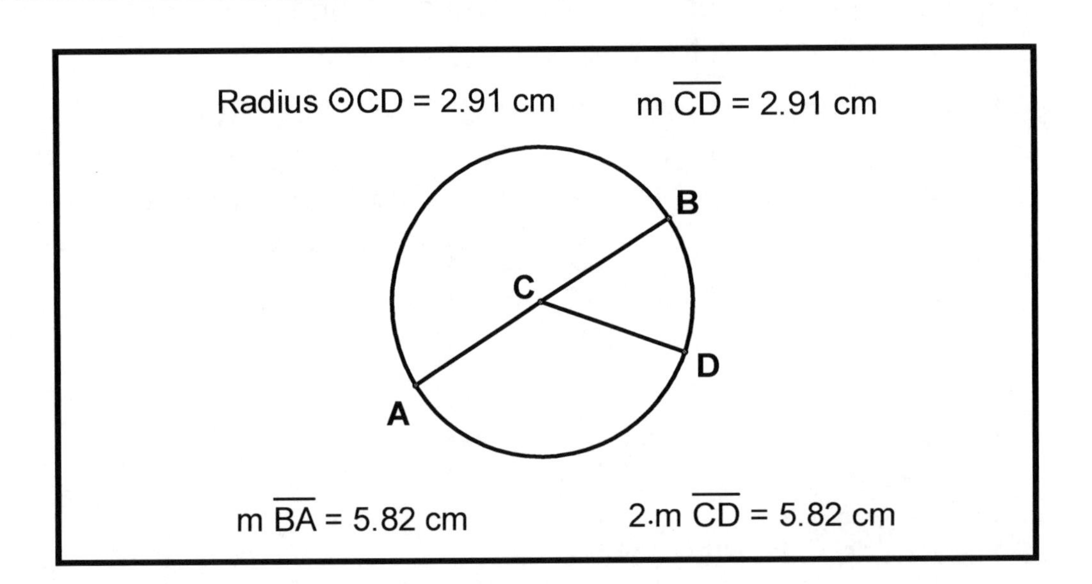

SUGGESTED EXERCISES

Type the solutions to the following problems in the blank region of your sketch.

1. Find the diameter of a circle whose radius is 4 in.

2. Find the radius of the circle whose diameter is 12 in.

3. Find in terms of x, the radius of a circle whose diameter is 6x.

4. Find in terms of x, the diameter of a circle whose radius is 4x + 3.

LAB #59: THE CIRCUMFERENCE OF A CIRCLE AND π

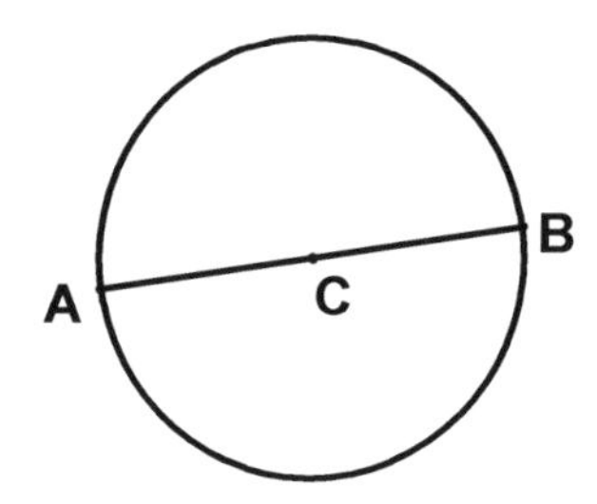

1. Select: the Point tool and plot two points about two inches apart from each other in the middle of your screen.
2. Select: the Arrow tool, [EDIT], [SELECT ALL], [CONSTRUCT], [SEGMENT], [MEASURE], [LENGTH], [EDIT], [SELECT PARENTS], [CONSTRUCT], [MIDPOINT], [DISPLAY], [SHOW LABEL] and deselect.
3. Select: points C, then B, [CONSTRUCT], [CIRCLE BY CENTER + POINT], [MEASURE], [CIRCUMFERENCE], [EDIT], [SELECT PARENTS], [MEASURE], [RADIUS] and deselect.
4. Select: [NUMBER], [CALCULATE], click on the caption that shows the measure of the circumference, [$\div$], click on the caption that shows the measure of diameter $\overline{AB}$, [OK] and deselect.
5. Click on point B, and keeping the left button depressed, drag it. Observe the number that remains the same as you do so.
6. Select: the Text tool, double click in the blank region to open a dialog box and (1) write the number that stayed the same as you changed the size of the circle, (2) name two parts of a circle that you have to divide to find that number.
7. Select: the Arrow tool, [NUMBER], [CALCULATE], [3.1416], [$*$], click on the caption that shows the measure of diameter $\overline{AB}$, [OK] and deselect.
8. Click on point B, and keeping the left button depressed, drag it. Observe the values that remain equal as you do so.
9. Select: the Text tool, double click in the blank region to open a dialog box and (1) describe your observations about the value of $3.1416 \cdot m\overline{AB}$ and the measure of the circumference of a circle,

(2) explain what number you have to multiply the diameter of the circle by to obtain the measure of its circumference.

10. Select: the Arrow tool, [NUMBER], [CALCULATE], [VALUE], [π], [*], click on the caption that shows the measure of diameter $\overline{AB}$, [OK] and deselect.

11. Select: [NUMBER], [CALCULATE], [2], [*], [VALUE], [π], [*], click on the caption that shows the measure of the radius of the circle and [OK].

12. Select: the Text tool, double click in the blank region to open a dialog box and (1) describe your observations about the values of $3.1416\cdot m\overline{AB}$, $\pi\cdot m\overline{AB}$, and $2\cdot\pi\cdot$ (Radius), (2) based on your previous answer, explain to what the value of π is equal, (3) express the diameter $\overline{AB}$ in terms of the radius of the circle, (4) give your own definition of the circumference of a circle.

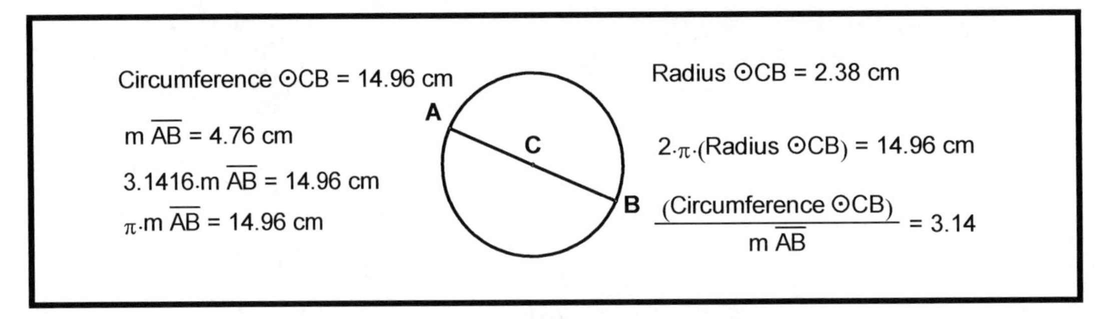

SUGGESTED EXERCISES

Type the solutions to the following problems in the blank region of your sketch.

1. Find the circumference of a circle whose diameter is 7 in (a) to the nearest tenth, (b) in terms of π.

2. Find the diameter of a circle whose circumference is 34 cm (a) to the nearest tenth, (b) in terms of π.

3. How many inches (to the nearest tenth of an inch) does the wheel of the scooter go in one turn if the diameter of its wheel is 9 inches?

4. The circumference of the circle is expressed by the formula $C = \pi d$, where d is the diameter of the circle. If the circumference is 5x, express the diameter d in terms of x and π.

LAB #60: THE AREA OF A CIRCLE

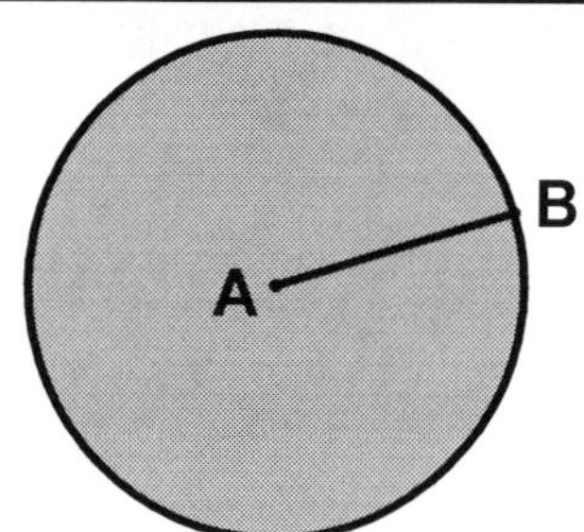

1. Select: the Point tool and plot two points approximately one inch apart from each other in the middle of your screen.
2. Select: the Arrow tool, [EDIT], [SELECT ALL], [CONSTRUCT], [SEGMENT], [MEASURE], [LENGTH], and deselect.
3. Select: points A, then B, [CONSTRUCT], [CIRCLE BY CENTER + POINT], [MEASURE] and [RADIUS].
4. Select: the Text tool, double click in the blank region to open a dialog box and describe your observations about the measure of line segment $\overline{AB}$ and the measure of the radius of the circle.
5. Select: the Arrow tool, click on point B and, keeping the left button depressed, drag it. Observe the values that remain equal as you do so.
6. Select: the Text tool, double click in the blank region to open a dialog box and explain what two points a line segment must connect to be called a radius of a circle.
7. Select: the Arrow tool and click in the blank region to deselect.
8. Select: the circle, [CONSTRUCT], [CIRCLE INTERIOR], [MEASURE], [AREA] and deselect.
9. Select: [NUMBER], [CALCULATE], [VALUE], [π], [$*$], click on the caption that shows the measure of the radius of the circle, [$\wedge$], [2] and [OK].
10. Select: the Text tool, double click in the blank region to open a dialog box and describe your observations about the value of ($\pi \cdot \text{Radius}^2$) and the measure of the area of a circle.

11. Select: the Arrow tool, [NUMBER], [CALCULATE], [3.1416], [∗], click on the caption that shows the measure of the radius of the circle, [∧], [2] and [OK].
12. Select: the Text tool, double click in the blank region to open a dialog box and (1) describe your observations about the value of ($3.1416 \cdot Radius^2$) and the value of ($\pi \cdot Radius^2$), (2) based on you previous answer; explain to what number the value of π is equal.
13. Select: the Arrow tool, click on point B, and keeping the left button depressed, drag it. Observe the values that remain equal as you do so.
14. Select: the Text tool, double click in the blank region to open a dialog box and explain what you have to do with the number π and the radius of the circle to calculate the area of the circle.

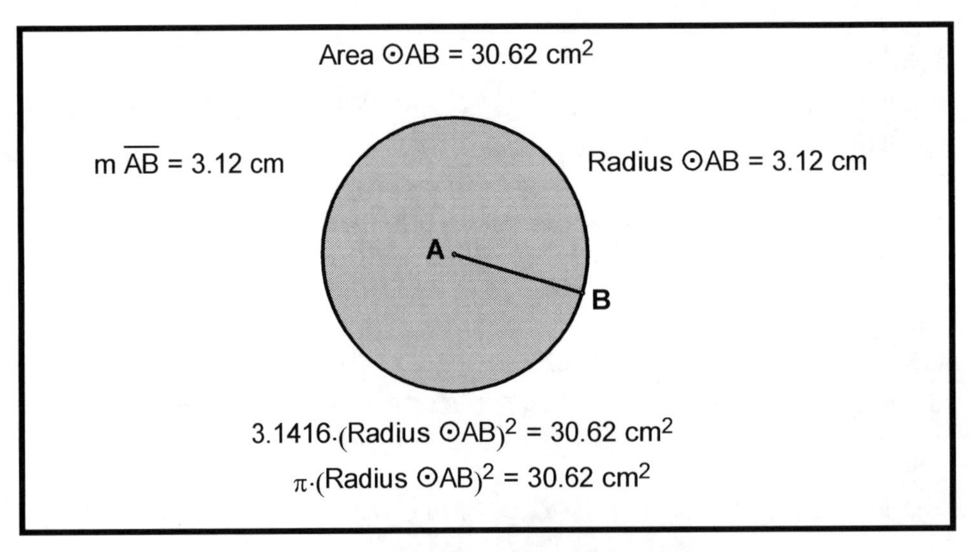

SUGGESTED EXERCISES

Type the solutions to the following problems in the blank region of your sketch.

1. Find in terms of π the area of a circle whose radius is 3 in.
2. Find the radius of the circle whose area is 153.938 cm^2.
3. If the area of a circle is 36π, find the radius.
4. If the diameter of a circle is 24 in, express in terms of π the area of a circle.

LAB #61: THE THREE ANGLE BISECTORS OF A TRIANGLE

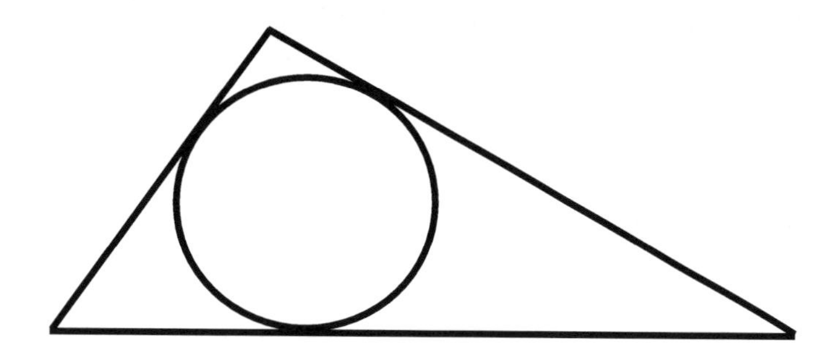

1. Select: [EDIT], [PREFERENCES], [TEXT], check the box [FOR ALL NEW POINTS] and click [OK].
2. Select: the Point tool and plot three points anywhere as the vertices of a triangle.
3. Select: the Arrow tool, [EDIT], [SELECT ALL], [CONSTRUCT], [SEGMENTS] and deselect.
4. Select: points A, then B, then C, [CONSTRUCT], [ANGLE BISECTOR] and deselect.
5. Select: points B, then A, then C, [CONSTRUCT], [ANGLE BISECTOR] and deselect.
6. Select: points A, then C, then B, [CONSTRUCT], [ANGLE BISECTOR] and deselect
7. Select: any two angle bisectors, [CONSTRUCT], [INTERSECTION], and deselect.
8. Select: all three angle bisectors, [DISPLAY] and [HIDE BISECTORS].
9. Select: point D, line segment $\overline{AB}$, [CONSTRUCT], [PERPENDICULAR LINE] and deselect.
10. Select: line segment $\overline{AB}$, perpendicular line passing through point D, [CONSTRUCT], [INTERSECTION] and deselect
11. Select: line $\overleftrightarrow{DE}$, [DISPLAY] and [HIDE PERPENDICULAR LINE].
12. Select: points D, E, [CONSTRUCT], [SEGMENT], [MEASURE], [LENGTH] and deselect.
13. Select: points D, then E, [CONSTRUCT], [CIRCLE BY CENTER + POINT], [MEASURE], [RADIUS] and deselect.
14. Select: point D, line segment $\overline{AB}$, [MEASURE], [DISTANCE] and deselect.
15. Select: point D, line segment $\overline{AC}$, [MEASURE], [DISTANCE] and deselect.

16. Select: point D, line segment $\overline{BC}$, [MEASURE] and [DISTANCE].
17. Select: the Text tool, double click in the blank region to open a dialog box and describe your observations about the measure of the radius of the circle and the distance from the point of intersection D (created by the angle bisectors of the triangle) to the sides $\overline{AB}$, $\overline{BC}$, and $\overline{AC}$ of ΔABC.
18. Select: the Arrow tool, click on any labeled point, and keeping the left button depressed, drag it. Observe the values that remain equal as you do so.
19. Select: the Text tool, double click in the blank region to open a dialog box and explain why point D is called the incenter of ΔABC.

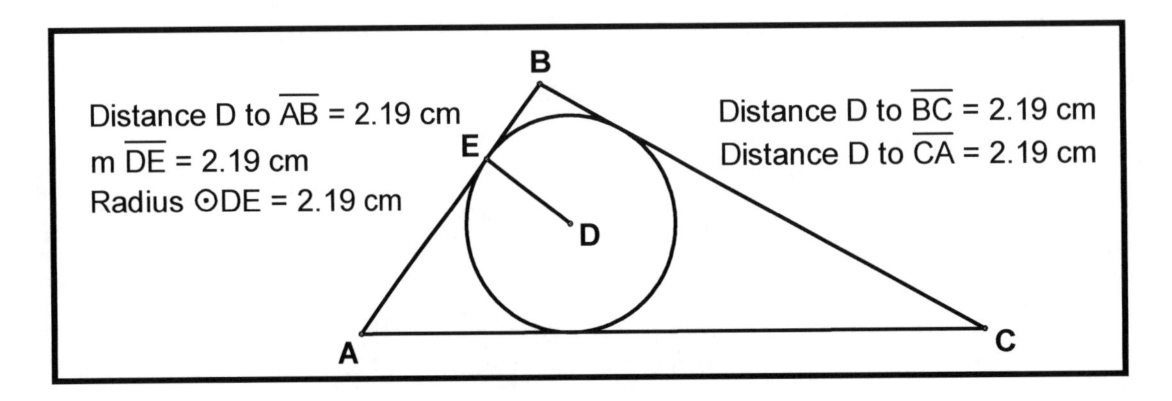

SUGGESTED EXERCISES

Type the solutions to the following problems in the blank region of your sketch.

1. Point D is the center of the circle inscribed into ΔABC. If the measure of the radius $\overline{DE}$ of the circle is 7 cm, find the shortest distance from incenter D to line segment $\overline{AC}$.

2. Point D is the center of the circle inscribed into ΔABC. If the measure of the radius $\overline{DE}$ of the circle is 4x and the distance from incenter D to line segment $\overline{AB}$ is 32 in, find the value of x.

3. Point D is the center of the circle inscribed into ΔABC. If the measure of the radius $\overline{DE}$ of the circle is 5x + 12 and the distance from the incenter D to line segment $\overline{BC}$ is 2x + 24, find the value of x and the length of radius $\overline{DE}$.

LAB #62: THE TANGENT RATIO IN A RIGHT TRIANGLE

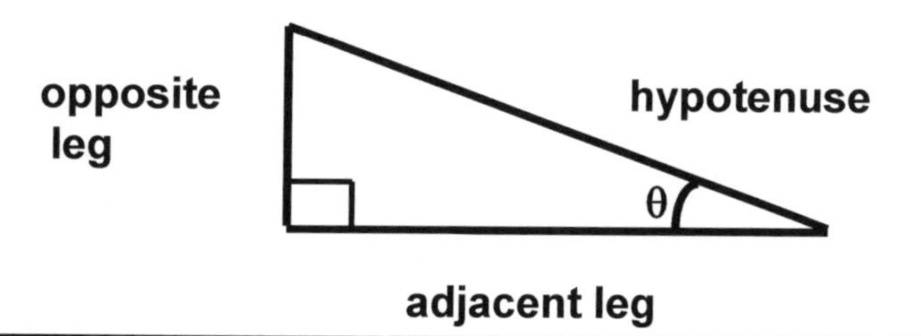

1. Select: [EDIT], [PREFERENCES], [TEXT], check the box [FOR ALL NEW POINTS] and click [OK].
2. Select: the Point tool and plot two points as the endpoints of a horizontal line segment.
3. Select: the Arrow tool, [EDIT], [SELECT ALL], [CONSTRUCT], [SEGMENT], [CONSTRUCT], [POINT ON SEGMENT] and deselect.
4. Select: points B, C, line segment $\overline{AB}$, [CONSTRUCT], [PERPENDICULAR LINES] and deselect.
5. Select: the perpendicular line passing through point B, [CONSTRUCT] and [POINT ON PERPENDICULAR LINE].
6. Click on point D, and keeping the left button depressed, drag it approximately one inch above point B and deselect.
7. Select: points A, D, [CONSTRUCT], [SEGMENT] and deselect.
8. Select: line segment $\overline{AD}$, the perpendicular line passing through C, [CONSTRUCT], [INTERSECTION], and deselect.
9. Select: lines $\overleftrightarrow{CE}$, $\overleftrightarrow{BD}$, [DISPLAY] and [HIDE PERPENDICULAR LINES].
10. Select: points A, then B, then D, [CONSTRUCT], [SEGMENTS], [MEASURE], [LENGTHS] and deselect.
11. Select: points A, then C, then E, [CONSTRUCT], [SEGMENTS], [DISPLAY], [COLOR], click on the red bar, [MEASURE], [LENGTHS] and deselect.
12. Select: dashed sides $\overline{AB}$ and $\overline{AD}$, [DISPLAY], [LINE STYLE], [SOLID] and deselect.
13. Select: points B, then A, then D, [MEASURE], [ANGLE] and deselect.
14. Select: [NUMBER], [CALCULATE], click on the caption that shows $m\overline{BD}$, [÷], click on the caption that shows $m\overline{BA}$, [OK] and deselect.
15. Repeat step 11 to find the quotient or ratio of the line segments $\overline{CE}$ to $\overline{AC}$.
16. Select: [NUMBER], [CALCULATE], [FUNCTIONS], [TAN], click on the caption that shows $m\angle BAD$, [OK] and deselect.

17. Select: the Text tool, double click in the blank region to open a dialog box and (1) describe your observations of the value of tan∠BAD and the values of $\frac{m\overline{BD}}{m\overline{BA}}$ and $\frac{m\overline{CE}}{m\overline{AC}}$, (2) use your observations to replace the question mark in the expressions tan∠BAD **?** $\frac{m\overline{BD}}{m\overline{BA}}$ and tan∠BAD **?** $\frac{m\overline{CE}}{m\overline{AC}}$ with the appropriate sign, >, < or =, (3) explain which two line segments of the right triangle (leg opposite the angle, leg adjacent to the angle or hypotenuse) must be divided by to find the tangent of an angle.
18. Select: the Arrow tool, click on point C, and keeping the left button depressed, drag it. Observe the values of the sides $\overline{CE}$ and $\overline{AC}$, ∠BAD, and tan∠BAD as you do so.
19. Select: the Text tool, double click in the blank region to open a dialog box and write if the value of the tangent depends on the measure of the sides $\overline{CE}$ and $\overline{AC}$.
20. Select: the Arrow tool, click on point D, and drag it up or down. Observe the values of ∠BAD, and tan∠BAD as you do so.
21. Select: the Text tool, double click in the blank region to open a dialog box and explain if the value of the tangent depends on the measure of the angle.

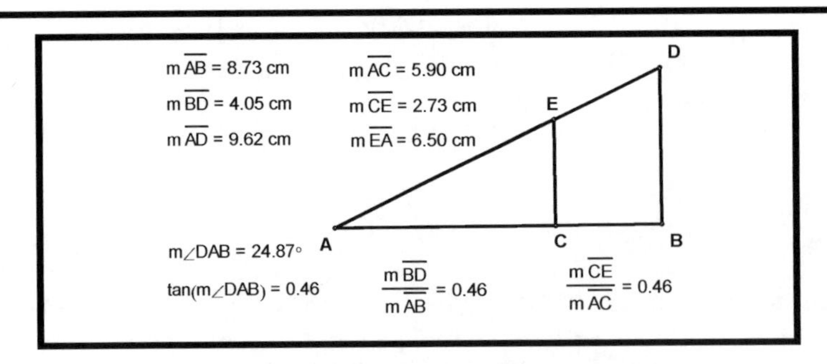

SUGGESTED EXERCISES

Type the solutions to the following problems in the blank region of your sketch.

1. In right ΔABD, ∠B is the right triangle. If m∠DAB = 20°, m$\overline{BD}$ = 2.6 cm, and m$\overline{AB}$ = 7.2 cm, find the tangent of ∠DAB to the nearest hundredth.
2. In right ΔABD, ∠B is the right angle. If m$\overline{AB}$ = 6 and the tangent of ∠DAB is 1.75, find the length of $\overline{BD}$ to the nearest hundredth.
3. In right ΔABD, ∠B is the right triangle. Find the length of $\overline{AB}$ to the nearest hundredth if m$\overline{BD}$ = 4 and the tangent of ∠DAB is 2.68.

LAB #63: THE SINE RATIO IN A RIGHT TRIANGLE

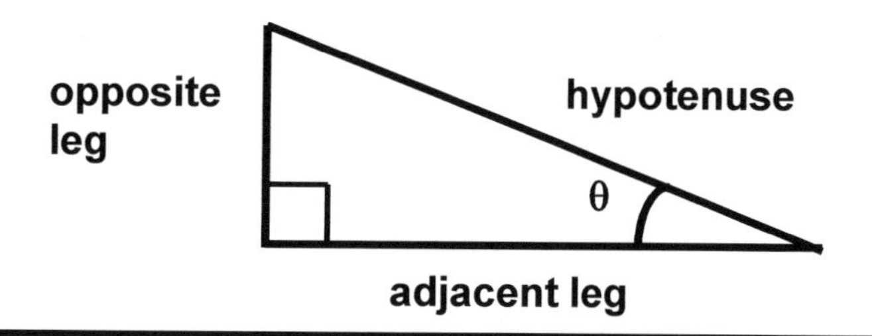

1. Select: [EDIT], [PREFERENCES], [TEXT], check the box [FOR ALL NEW POINTS] and click [OK].
2. Select: the Point tool and plot two points as the endpoints of a horizontal line segment.
3. Select: the Arrow tool, [EDIT], [SELECT ALL], [CONSTRUCT], [SEGMENT], [CONSTRUCT], [POINT ON SEGMENT] and deselect.
4. Select: points B, C, line segment $\overline{AB}$, [CONSTRUCT], [PERPENDICULAR LINES] and deselect.
5. Select: the perpendicular line passing through B, [CONSTRUCT] and [POINT ON PERPENDICULAR LINE].
6. Click on point D, and keeping the left button depressed, drag it approximately one inch above point B and deselect.
7. Select: points A, D, [CONSTRUCT], [SEGMENT] and deselect.
8. Select: line segment $\overline{AD}$, the perpendicular line passing through C, [CONSTRUCT], [INTERSECTION], and deselect.
9. Select: lines $\overleftrightarrow{CE}$, $\overleftrightarrow{BD}$, [DISPLAY] and [HIDE PERPENDICULAR LINES].
10. Select: points A, then B, then D, [CONSTRUCT], [SEGMENTS], [MEASURE], [LENGTHS] and deselect.
11. Select: points A, then C, then E, [CONSTRUCT], [SEGMENTS], [DISPLAY], [COLOR], click on the red bar, [MEASURE], [LENGTHS] and deselect.
12. Select: dashed sides $\overline{AB}$ and $\overline{AD}$, [DISPLAY], [LINE STYLE], [SOLID] and deselect.
13. Select: points B, then A, then D, [MEASURE], [ANGLE] and deselect.
14. Select: [NUMBER], [CALCULATE], click on the caption that shows m$\overline{BD}$, [÷], click on the caption that shows m$\overline{AD}$, [OK] and deselect.
15. Repeat step 11 to find the quotient, or ratio of line segments $\overline{CE}$ to $\overline{EA}$.
16. Select: [NUMBER], [CALCULATE], [FUNCTIONS], [SIN], click on the caption that shows m∠BAD and [OK].

17. Select: the Text tool, double click in the blank region to open a dialog box and (1) describe your observations of the value of sin∠BAD and the values of $\frac{m\overline{BD}}{m\overline{AD}}$ and $\frac{m\overline{CE}}{m\overline{EA}}$, (2) use your observations to replace the question mark in the expressions sin∠BAD **?** $\frac{m\overline{BD}}{m\overline{AD}}$ and sin∠BAD **?** $\frac{m\overline{CE}}{m\overline{EA}}$ with the appropriate sign, >, <, or =, (3) explain which two line segments of the right triangle (leg opposite to the angle, leg adjacent to the angle, or hypotenuse) must be divided to find the sine of an angle.
18. Select: the Arrow tool, click on point C, and keeping the left button depressed, drag it. Observe the values of ∠BAD, sin∠BAD, sides $\overline{CE}$ and $\overline{EA}$ as you do so.
19. Select: the Text tool, double click in the blank region to open a dialog box and write if the value of sin∠BAD depends on the measure of the sides $\overline{CE}$ and $\overline{EA}$.
20. Select: the Arrow tool, click on point D and drag it up or down. Observe the measure of ∠BAD and the value of sin∠BAD.
21. Select: the Text tool, double click in the blank region to open a dialog box and explain if the value of sin∠BAD depends on the measure of ∠BAD.

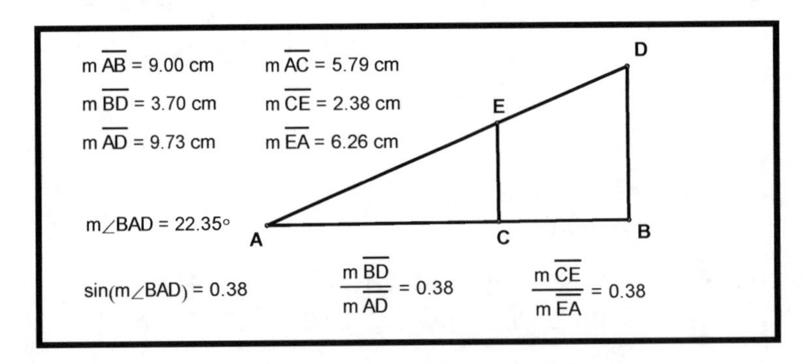

SUGGESTED EXERCISES

Type the solutions to the following problems in the blank region of your sketch.

1. In Δ ABD, ∠B is a right angle. If m∠BAD = 56°, m$\overline{BD}$ = 5.99 cm, and m$\overline{AD}$ = 7.224 cm, find sin∠BAD to the nearest hundredth.

2. In Δ ABD, ∠B is a right angle. Find the measure of side $\overline{BD}$ to the nearest thousandth if m$\overline{AD}$ = 7 and sin∠BAD = 0.7564.

3. Δ ACE is a right triangle. Find the measure of hypotenuse $\overline{EA}$ if m$\overline{CE}$ = 5 cm, and sin∠CAE = 0.8547. Round your answer to the nearest hundredth.

LAB #64: THE COSINE RATIO IN A RIGHT TRIANGLE

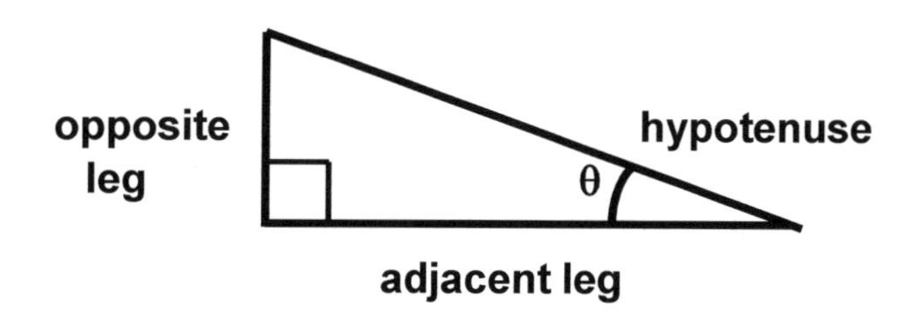

1. Select: [EDIT], [PREFERENCES], [TEXT], check the box [FOR ALL NEW POINTS] and click [OK].
2. Select: the Point tool and plot two points as the endpoints of a horizontal line segment.
3. Select: the Arrow tool, [EDIT], [SELECT ALL], [CONSTRUCT], [SEGMENT], [CONSTRUCT], [POINT ON SEGMENT] and deselect.
4. Select: points B, C, line segment $\overline{AB}$, [CONSTRUCT], [PERPENDICULAR LINES] and deselect.
5. Select: the perpendicular line passing through point B, [CONSTRUCT] and [POINT ON PERPENDICULAR LINE].
6. Click on point D, and keeping the left button depressed, drag it approximately one inch above point B and deselect.
7. Select: points A, D, [CONSTRUCT], [SEGMENT] and deselect.
8. Select: line segment $\overline{AD}$, the perpendicular line passing through C, [CONSTRUCT], [INTERSECTION], and deselect.
9. Select: lines $\overleftrightarrow{CE}$, $\overleftrightarrow{BD}$, [DISPLAY] and [HIDE PERPENDICULAR LINES].
10. Select: points A, then B, then D, [CONSTRUCT], [SEGMENTS], [MEASURE], [LENGTHS] and deselect.
11. Select: points A, then C, then E, [CONSTRUCT], [SEGMENTS], [DISPLAY], [COLOR], click on the red bar, [MEASURE], [LENGTHS] and deselect.
12. Select: dashed sides $\overline{AB}$ and $\overline{AD}$, [DISPLAY], [LINE STYLE], [SOLID] and deselect.
13. Select: points B, then A, then D, [MEASURE], [ANGLE] and deselect.
14. Select: [NUMBER], [CALCULATE], click on the caption that shows $m\overline{BA}$, [÷], click on the caption that shows $m\overline{AD}$, [OK] and deselect.
15. Repeat step 11 to find the quotient or ratio of line segments $\overline{AC}$ to $\overline{EA}$.
16. Select: [NUMBER], [CALCULATE], [FUNCTIONS], [COS], click on the caption that shows $m\angle BAD$ and [OK].

17. Select: the Text tool, double click in the blank region to open a dialog box and (1) describe your observations of the value of cos∠BAD and the values of $\frac{m\overline{BA}}{m\overline{AD}}$ and $\frac{m\overline{AC}}{m\overline{EA}}$, (2) use your observations to replace the question mark in the expressions cos∠BAD **?** $\frac{m\overline{BA}}{m\overline{AD}}$ and cos∠BAD **?** $\frac{m\overline{AC}}{m\overline{EA}}$ with the appropriate sign, >, <, or =, (3) explain which two line segments of the right triangle (leg opposite to the angle, leg adjacent to the angle or hypotenuse) must be divided to find the cosine of an angle.
18. Select: the Arrow tool, click on point C, and keeping the button depressed, drag it. Observe the values of ∠BAD, cos∠BAD, sides $\overline{AC}$ and $\overline{EA}$.
19. Select: the Text tool, double click in the blank region to open a dialog box and write if the value of cos∠BAD depends on the measures of the sides $\overline{AC}$ and $\overline{EA}$.
20. Select: the Arrow tool, click on point D, and drag it up or down. Observe the measure of ∠BAD and the value of cos∠BAD.
21. Select: the Text tool, double click in the blank region to open a dialog box and explain if the value of cos∠BAD depends on the measure of ∠BAD.

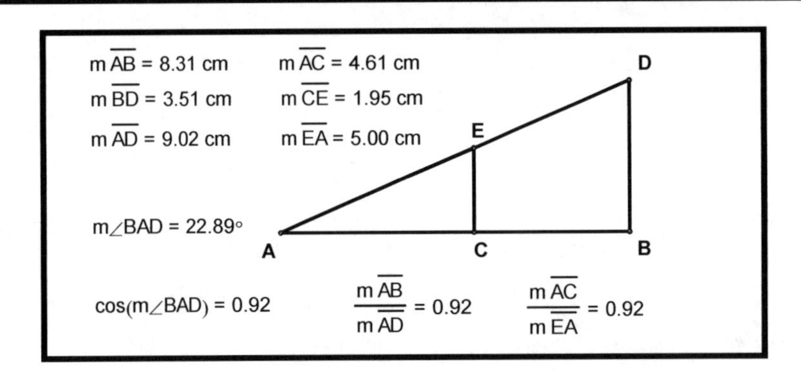

SUGGESTED EXERCISES

Type the solutions to the following problems in the blank region of your sketch.

1. Δ ABD is a right triangle. If m∠BAD = 54°, m$\overline{AB}$= 3.75 cm, and m$\overline{AD}$ = 6.38 cm, find cos(∠BAD) to the nearest thousandth.

2. Δ ABD is a right triangle. Find the measure of side $\overline{AB}$ to the nearest hundredth, if m$\overline{AD}$ = 9 and cos(∠BAD) = 0.3456.

3. Δ ACE is a right triangle. Find the measure of hypotenuse $\overline{EA}$ if m$\overline{AC}$ = 4.76 cm, and cos(∠CAE) = 0.6801. Round your answer to the nearest integer.

LAB #65: NAMING A POINT BY USING ITS COORDINATES

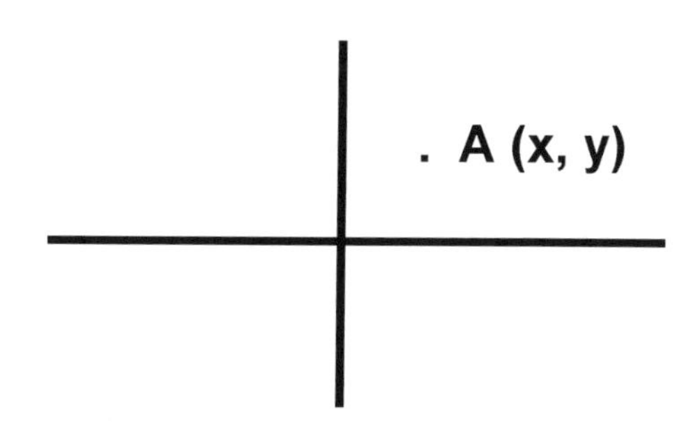

1. Select: the Point tool and plot a point.
2. Select: the Arrow tool, [EDIT], [SELECT ALL], [MEASURE], [COORDINATES], [GRAPH], [SNAP POINTS] and deselect.
3. Select: point A, [MEASURE] and [ABSCISSA (x)].
4. Select: the Text tool, double click to open a dialog box and describe your observations about the value of the first number in the pair of the coordinates, you have measured, and the value of the x-coordinate of the point A.
5. Select: the Arrow tool and click in the blank region to deselect.
6. Select: point A, [MEASURE] and [ORDINATE (y)].
7. Select: the Text tool, double click to open a dialog box and describe your observations about the value of the second number in the pair of the coordinates that you have just measured, and the value of the y-coordinate of the point A.
8. Select: the Arrow tool, click on the point A, and keeping the left button depressed, drag it. Observe the values that remain equal as you do so.
9. Select: the Text tool, double click in the blank region to open a dialog box, and explain (1) which coordinate (x or y) is represented by the first number in the pair of the coordinates of the point, (2) which coordinate (x or y) is represented by the second number in the pair of the coordinates of the point.
10. Select: the Arrow tool, point A and keeping the left button depressed, drag it to the 1st quadrant (the upper right corner of the coordinate plane).
11. Select: the Text tool, double click in the blank region to open a dialog box and describe if both coordinates have a positive or a negative value.
12. Select: the Arrow tool, point A, and keeping the left button depressed, drag it to the 2nd quadrant (the upper left part of the coordinate plane).

13. Select: the Text tool, double click in the blank region to open a dialog box and (1) name which coordinate (x or y) has a positive value (2) name which coordinate (x or y) has a negative value.
14. Select: the Arrow tool, point A, and keeping the left button depressed, drag it to the 3rd quadrant (the lower left corner of the coordinate plane).
15. Select: the Text tool, double click in the blank region to open a dialog box and describe if both coordinates have a positive or a negative value.
16. Select: the Arrow tool, point A, and keeping the left button depressed, drag it to the 4th quadrant (the lower right corner of the coordinate plane).
17. Select: the Text tool, double click in the blank region to open a dialog box and (1) name which coordinate (x or y) has a positive value, (2) name which coordinate (x or y) has a negative value.

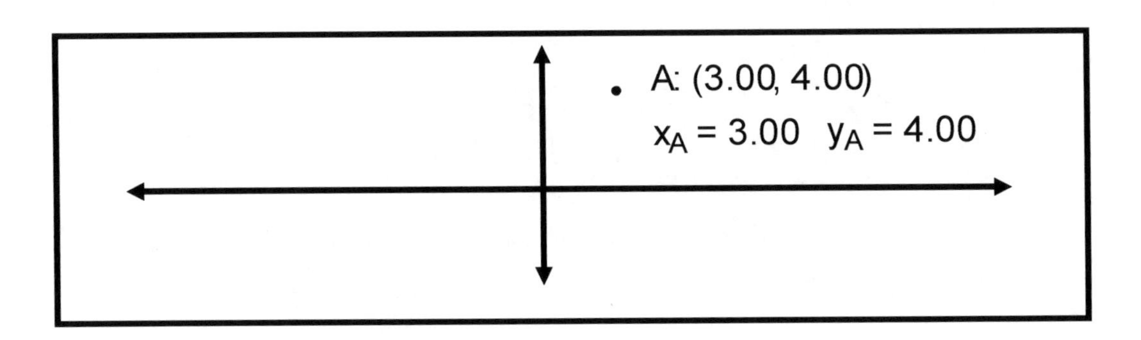

SUGGESTED EXERCISES

Type the solutions to the following problems in the blank region of your sketch.

1. Given point B (5, -7), name (1) the x-coordinate of the point B and (2) name the y-coordinate of the point B.

2. Plot a point on the coordinate plane whose coordinates are (3, 5). Label the point C.

3. Plot a point on the coordinate plane whose coordinates are (-4, 6). Label the point D.

4. Plot a point on the coordinate plane whose coordinates are (-5, -3). Label the point E.

LAB #66: THE SLOPE OF A LINE

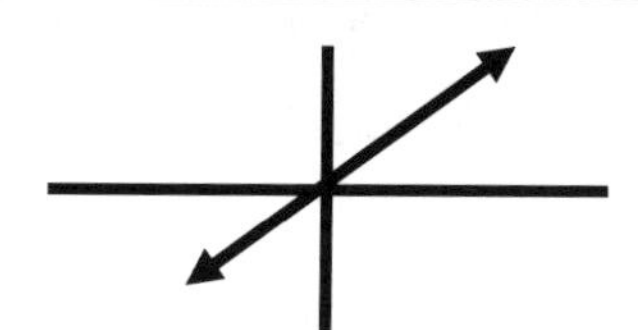

1. Select: [EDIT], [PREFERENCES], [TEXT], check the box [FOR ALL NEW POINTS] and click [OK].
2. Select: [GRAPH], [SHOW GRID], [GRAPH] and [SNAP POINTS].
3. Select: the Point tool and plot two points anywhere in the plane.
4. Select: the Arrow tool and click in the blank region to deselect.
5. Select: points A, B, [CONSTRUCT], [LINE], [MEASURE], [SLOPE] and deselect.
6. Select: points A, B, [MEASURE] and [ORDINATE (y)].
7. Select: [NUMBER], [CALCULATE], click on the caption that shows the value of y_B, [–], click on the caption that shows the value of y_A, [OK] and deselect.
8. Select: points A, B, [MEASURE] and [ABSCISSAE (x)].
9. Select: [NUMBER], [CALCULATE], click on the caption that shows the value of x_B, [–], click on the caption that shows the value of x_A, [OK] and deselect.
10. Select: [NUMBER], [CALCULATE], click on the caption that shows the value of $(y_B - y_A)$, [÷], click on the caption that shows the value of $(x_B - x_A)$ and [OK].
11. Select: the Text tool, double click in the blank region to open a dialog box and describe your observations about the value of the slope of the line $\overleftrightarrow{AB}$ and the value of $\frac{y_B - y_A}{x_B - x_A}$.
12. Select: the Arrow tool, click on point A, and keeping the left button depressed, drag it. Do the same to drag point B. Observe the values that remain equal as you do so.
13. Select: the Text tool, double click in the blank region to open a dialog box and explain what you have to subtract and divide to find the slope of a line.
14. Select: the Arrow tool and click in the blank region to deselect.
15. Select: point A, the y-axis, [CONSTRUCT], [PERPENDICULAR LINE] and deselect.
16. Select: point B, the x-axis, [CONSTRUCT], [PERPENDICULAR LINE] and deselect.

17. Select: the perpendicular line passing through point A, the perpendicular line passing through point B, [CONSTRUCT], [INTERSECTION] and deselect.

18. Select: lines $\overleftrightarrow{AC}$, $\overleftrightarrow{BC}$, [DISPLAY] and [HIDE PERPENDICULAR LINES].

19. Select: points A, then C, then B, [CONSTRUCT], [SEGMENTS], [MEASURE], [LENGTHS] and deselect.

20. Select: the dashed line $\overleftrightarrow{AB}$, [DISPLAY], [LINE STYLE], [SOLID] and deselect

21. Select: the Text tool, double click in the blank region to open a dialog box and 1) describe your observations about the measure of vertical line segment $\overline{BC}$ and the absolute value of $(y_B - y_A)$, 2) describe your observations about the measure of horizontal line segment $\overline{AC}$ and the absolute value of $(x_B - x_A)$, 3) use your two previous answers to give your own definition of the slope of a line.

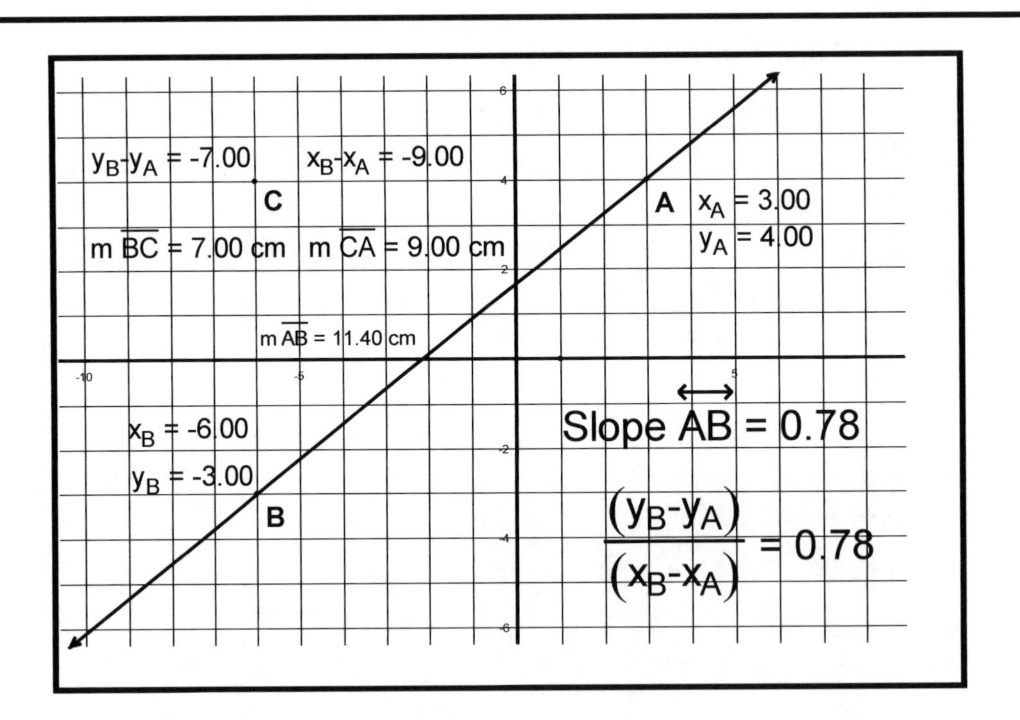

SUGGESTED EXERCISES

Type the solutions to the following problems in the blank region of your sketch.

1. Find the slope of the line passing through the points A (3, 4) and B (4, 6).

2. Find the slope of the line passing through the points C (-2, 3) and D (4, 5).

3. Find the slope of the line passing through the points E (-3, 4) and F (8, -7).

LAB #67: THE SLOPES OF PARALLEL LINES

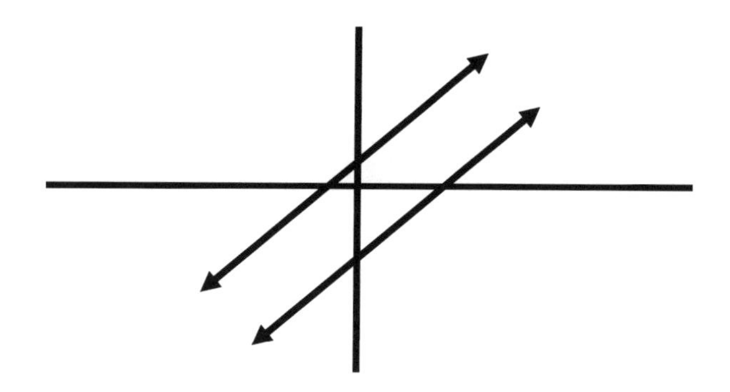

1. Select: the Point tool and plot two points randomly.
2. Select: the Arrow tool, [EDIT], [SELECT ALL], [DISPLAY], [SHOW LABELS], [CONSTRUCT], [LINE] and deselect.
3. Select: the Point tool and plot a point to the right of line $\overleftrightarrow{AB}$.
4. Select: the Arrow tool, [EDIT], [SELECT ALL], [EDIT], [SELECT CHILDREN], [CONSTRUCT], [PARALLEL LINE], [MEASURE], [SLOPE], [GRAPH], [SNAP POINTS] and deselect.
5. Select: line $\overleftrightarrow{AB}$, [MEASURE], [SLOPE] and deselect.
6. Click on point A or B, and keeping the left button depressed, drag it. Observe the measures that remain equal as you do so.
7. Select: the Text tool, double click in the blank region to open a dialog box and describe your observations about the slopes of two parallel lines.
8. Select: the Arrow tool and click in the blank region to deselect.
9. Select: line j, [TRANSFORM] and [TRANSLATE].
10. Type [2] in the displayed window marked as [FIXED DISTANCE] and click [TRANSLATE].
11. Select: [MEASURE], [SLOPE] and deselect.
12. Select: the Arrow tool, click on point A or B, and keeping the left button depressed, drag it. Observe the measures that remain equal as you do so.

13. Select: the Text tool, double click in the blank region to open a dialog box and (1) describe your observations about the slopes of lines $\overleftrightarrow{AB}$, j and j', (2) state whether or not line j', is parallel to the pair of parallel lines $\overleftrightarrow{AB}$ and j. Explain how the measures of the slopes of lines j', $\overleftrightarrow{AB}$ and j support your conclusion.

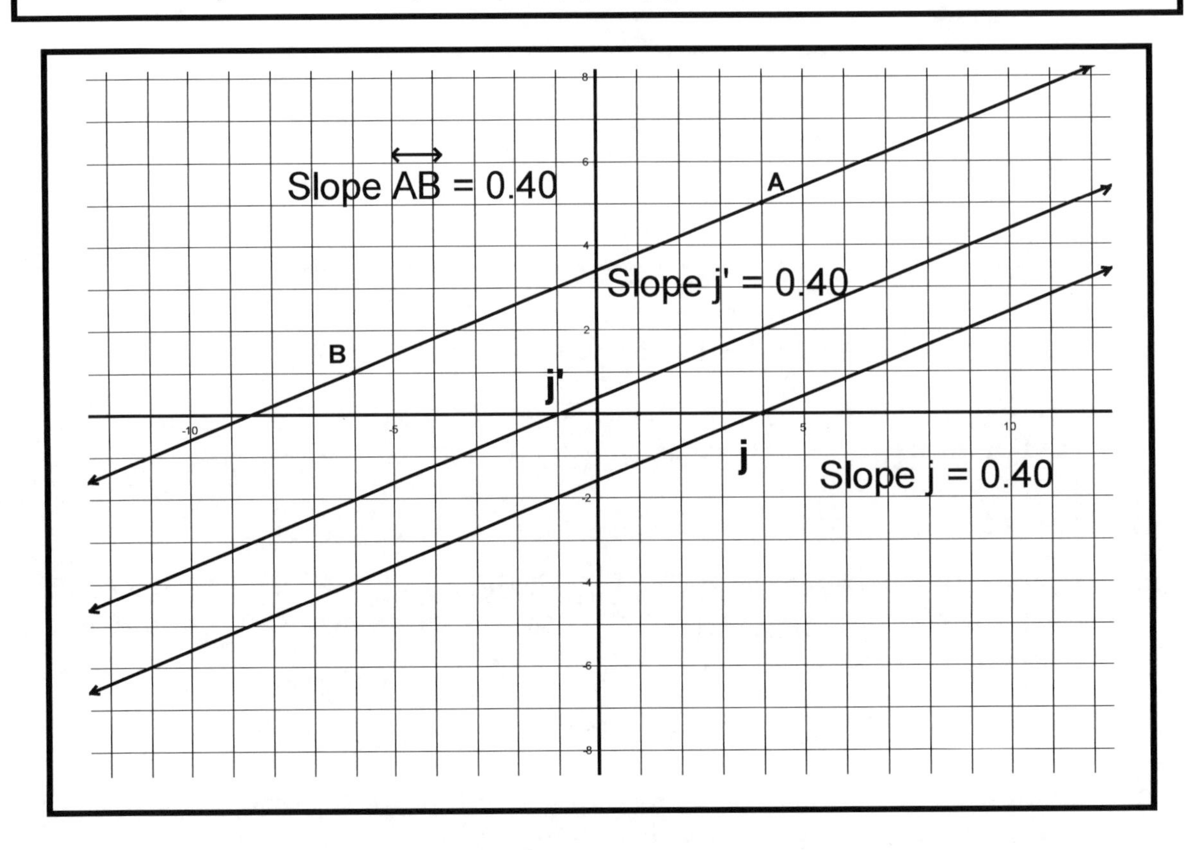

<u>SUGGESTED EXERCISES</u>

Type the solutions to the following problems in the blank region of your sketch.

1. Line $\overleftrightarrow{AB}$ is parallel to line $\overleftrightarrow{CD}$. Find the slope of line $\overleftrightarrow{CD}$ if the slope of line $\overleftrightarrow{AB}$ is 2.

2. Line $\overleftrightarrow{AB}$ is parallel to line $\overleftrightarrow{CD}$. If the slope of line $\overleftrightarrow{AB}$ is 5 and the slope of line $\overleftrightarrow{CD}$ is 2x, find the value of x.

3. Line $\overleftrightarrow{AB}$ is parallel to line $\overleftrightarrow{CD}$. If the slope of line $\overleftrightarrow{AB}$ is 2x + 2 and the slope of line $\overleftrightarrow{CD}$ is x + 4, find the value of x and each slope.

LAB #68: THE SLOPES OF PERPENDICULAR LINES

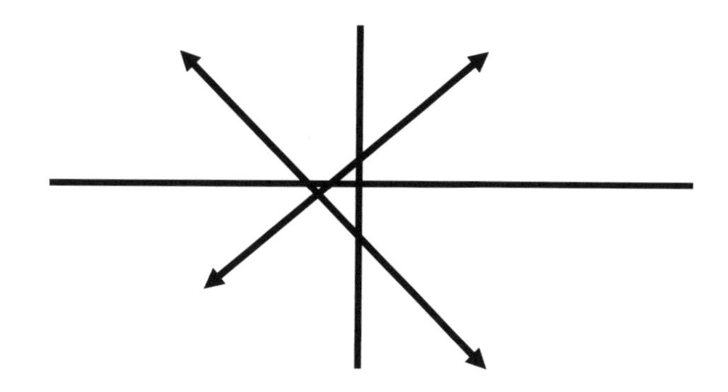

1. Select: the Point tool and plot two points randomly.
2. Select: the Arrow tool, [EDIT], [SELECT ALL], [DISPLAY], [SHOW LABELS], [CONSTRUCT], [LINE] and deselect.
3. Select: the Point tool and plot a point anywhere except on line $\overleftrightarrow{AB}$.
4. Select: the Arrow tool, [EDIT], [SELECT ALL], [EDIT], [SELECT CHILDREN], [CONSTRUCT], [PERPENDICULAR LINE], [MEASURE], [SLOPE], [GRAPH], [SNAP POINTS] and deselect.
5. Select: line $\overleftrightarrow{AB}$, [MEASURE], [SLOPE] and deselect.
6. Select: [NUMBER], [CALCULATE], click on the caption that shows slope $\overleftrightarrow{AB}$, [∗], click on the caption that shows slope j, [OK] and deselect.
7. Click on point A or B, and keeping the left button depressed, drag it. Observe the value that remains the same as you do so.
8. Select: the Text tool, double click in the blank region to open a dialog box and describe your observations about the product of the slopes of two perpendicular lines.
9. Select: the Arrow tool and click in the blank region to deselect.
10. Select: line j, [TRANSFORM] and [TRANSLATE].
11. Type [3] in the displayed window marked as [FIXED DISTANCE] and click [TRANSLATE].
12. Select: [MEASURE] and [SLOPE].

13. Select: [NUMBER], [CALCULATE], click on the caption that shows slope $\overleftrightarrow{AB}$, [∗], click on the caption that shows slope j', [OK] and deselect.
14. Click on point A or B, and keeping the left button depressed, drag it. Observe the values of the product of the slopes of lines $\overleftrightarrow{AB}$ and j, $\overleftrightarrow{AB}$ and j'.
15. Select: the Text tool, double click in the blank region to open a dialog box and (1) describe your observations about the product of the slopes of lines $\overleftrightarrow{AB}$ and j, $\overleftrightarrow{AB}$ and j', (2) state whether or not line j' is perpendicular to line $\overleftrightarrow{AB}$. Explain how the product of the slopes of lines $\overleftrightarrow{AB}$ and j, $\overleftrightarrow{AB}$ and j' support your conclusion.

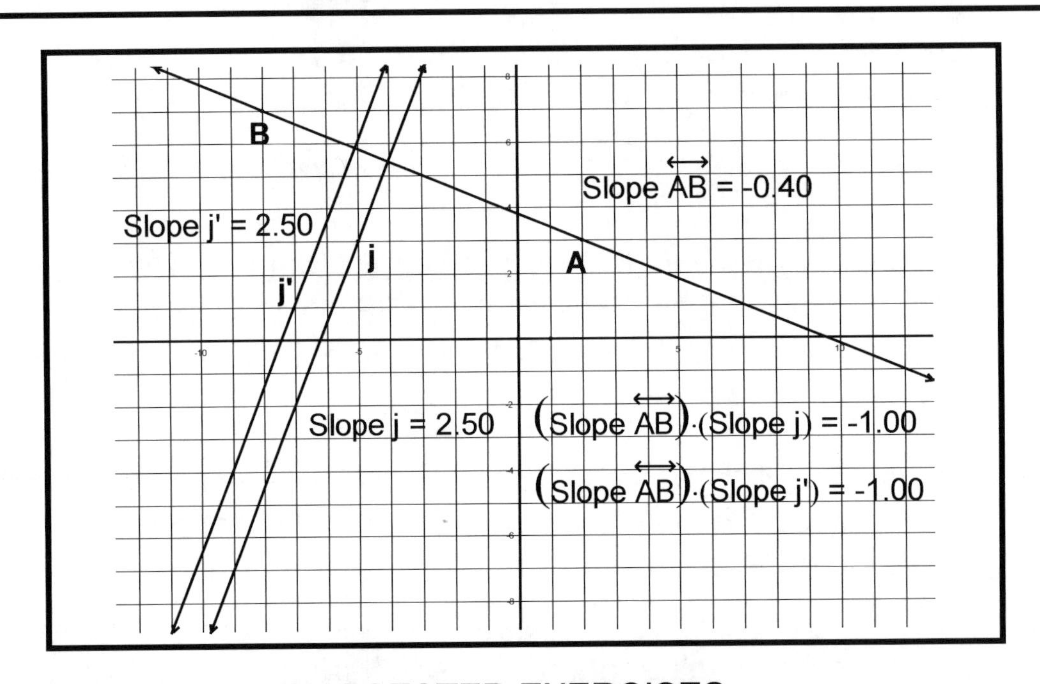

SUGGESTED EXERCISES

Type the solutions to the following problems in the blank region of your sketch.

1. The slope of line $\overleftrightarrow{AB}$ is 2. The slope of line $\overleftrightarrow{CD}$ is -0.5. Show that the lines are perpendicular.
2. Line $\overleftrightarrow{AB}$ is perpendicular to line $\overleftrightarrow{CD}$. If the slope of line $\overleftrightarrow{AB}$ is 4, find the slope of line $\overleftrightarrow{CD}$.
3. Line $\overleftrightarrow{AB}$ is perpendicular to line $\overleftrightarrow{CD}$. If the slope of line $\overleftrightarrow{AB}$ is x, express the slope of line $\overleftrightarrow{CD}$ in terms of x.

LAB #69: THE SLOPE AND THE Y-INTERCEPT OF A LINE

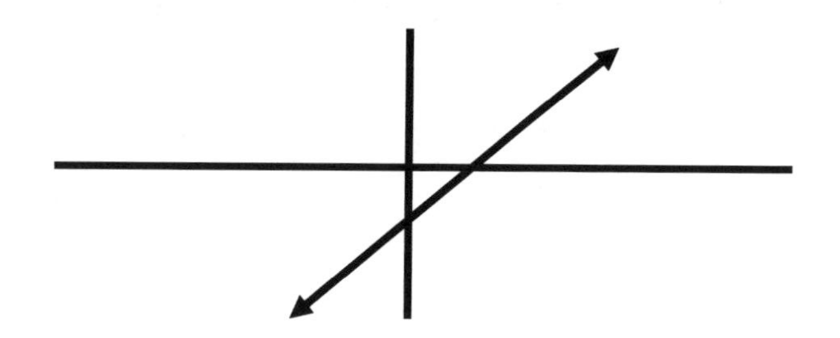

1. Select: [EDIT], [PREFERENCES], [TEXT], check the box [FOR ALL NEW POINTS] and click [OK].
2. Select: [GRAPH], [SHOW GRID], [GRAPH] and [SNAP POINTS].
3. Select: the y-axis, [CONSTRUCT], [POINT ON AXIS] and deselect.
4. Select: the Point tool and plot a point anywhere on the coordinate plane.
5. Select: the Arrow tool and click in the blank region to deselect.
6. Select: points A, B, [CONSTRUCT], [LINE], [MEASURE], [EQUATION] and deselect.
7. Select: point A, [MEASURE] and [ORDINATE (y)].
8. Select: the Text tool, double click in the blank region to open a dialog box and describe your observations about the value of the y-coordinate of the point A and the number that follows the x-term in the equation of the line $\overleftrightarrow{AB}$.
9. Select: the Arrow tool, click on point A, and keeping the left button depressed, drag it along y-axis. Observe the values that remain equal as you do so.
10. Select: the Text tool, double click in the blank region to open a dialog box and explain why the number that follows the x-term in the equation of the line is called the y-intercept.
11. Select: the Arrow tool and click in the blank region to deselect.
12. Select: line $\overleftrightarrow{AB}$, [MEASURE] and [SLOPE].
13. Select: the Text tool, double click in the blank region to open a dialog box and describe your observations about the value of the slope of line $\overleftrightarrow{AB}$ and the value of the coefficient of the x-term in the equation of line $\overleftrightarrow{AB}$.

14. Select: the Arrow tool, click on point B, and keeping the left button depressed, drag it up and down. Observe the values that remain equal as you do so.
15. Select: the Text tool, double click in the blank region to open a dialog box and explain (1) how the slope of the line is related to the coefficient of the x-term in the equation of the line, (2) what is represented by *m* in the general equation of the line y = mx + b, (3) what is represented by *b* in the general equation of the line y = mx + b

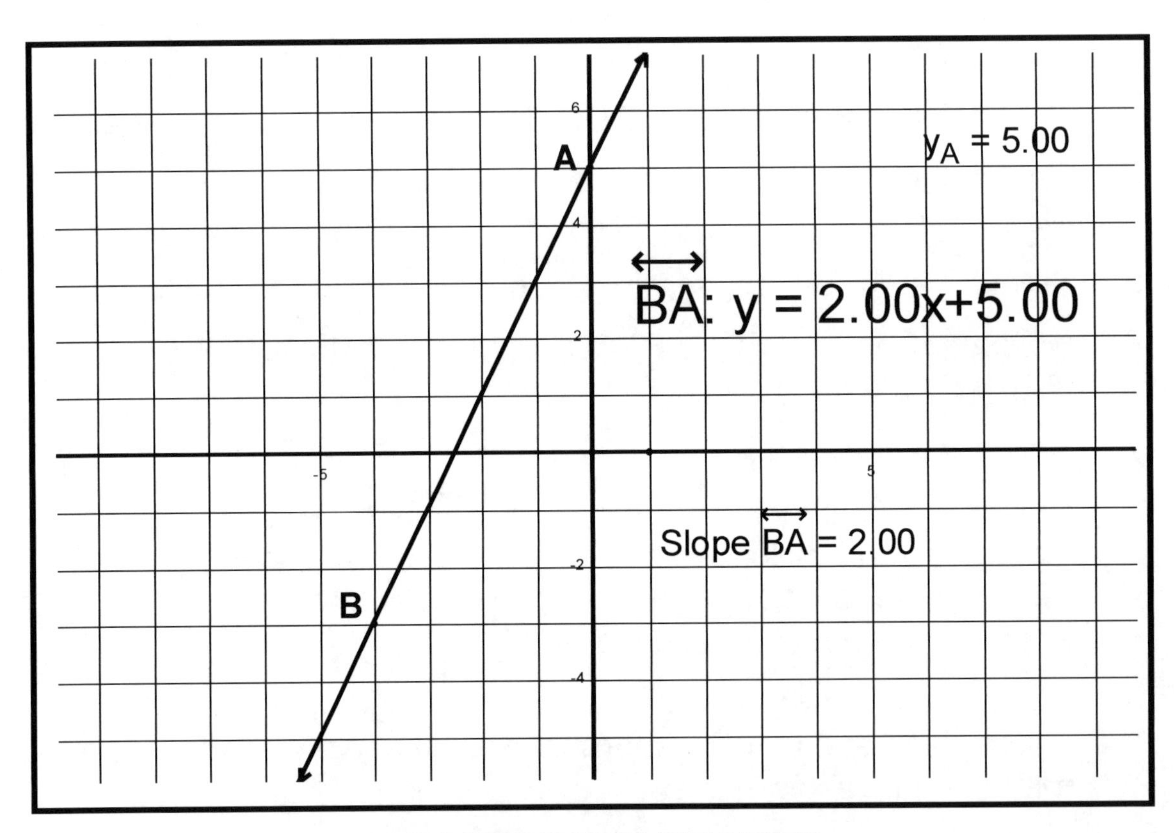

SUGGESTED EXERCISES

Type the solutions to the following problems in the blank region of your sketch.

1. Find the slope and y-intercept of the line given by the equation y = 2x + 4.
2. Find the slope and y-intercept of the line given by the equation y = 3x – 7.
3. Write the equation of the line whose slope is 4 and y-intercept is 6.
4. Write the equation of the line whose slope is 1 and y-intercept is -2.

LAB #70: THE EQUATION OF A VERTICAL LINE ON THE COORDINATE PLANE

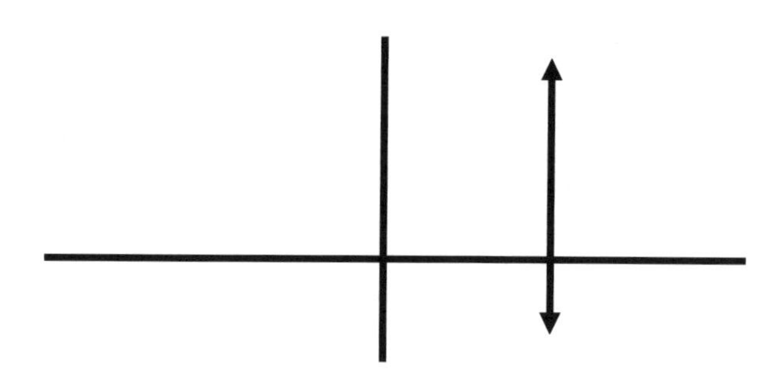

1. Select: [EDIT], [PREFERENCES], [TEXT], check the box [FOR ALL NEW POINTS] and click [OK].
2. Select: [GRAPH], [SHOW GRID], [GRAPH] and [SNAP POINTS].
3. Select: the Point tool and plot a point anywhere on the coordinate plane.
4. Select: the Arrow tool and click in the blank region to deselect.
5. Select: point A, the x-axis, [CONSTRUCT], [PERPENDICULAR LINE], [CONSTRUCT], [POINT ON PERPENDICULAR LINE] and deselect.
6. Click on point B, and keeping the left button depressed, drag it about two inches away from point A and deselect
7. Select: points A, B, [MEASURE], [ABSCISSAE (x)] and deselect.
8. Select: point A, B, [MEASURE], [ORDINATES (y)] and deselect.
9. Click on point B, and keeping the left button depressed, drag it along line $\overleftrightarrow{AB}$. Observe the values that remain the same as you do so.
10. Select: the Text tool, double click in the blank region to open a dialog box and describe your observations about the values of the x-coordinates of points A and B.
11. Select: the Arrow tool and click in the blank region to deselect.
12. Select: line $\overleftrightarrow{AB}$, [MEASURE] and [EQUATION]. The vertical line $\overleftrightarrow{AB}$ will be labeled as j, and the equation of the vertical line j will be displayed on your sketch.

13. Select: the Text tool, double click in the blank region to open a dialog box and (1) write the equation of the vertical line j, (2) explain why, you think, the equation of the vertical line is the same as the value of the x-coordinates of the points on this line.
14. Select: the Arrow tool, click on point A, and keeping the left button depressed, drag it to the right or to the left. Observe the values that remain equal as you do so.
15. Select: the Text tool, double click in the blank region to open a dialog box and explain how the equation of the vertical line is related to the x-coordinate of points A or B.

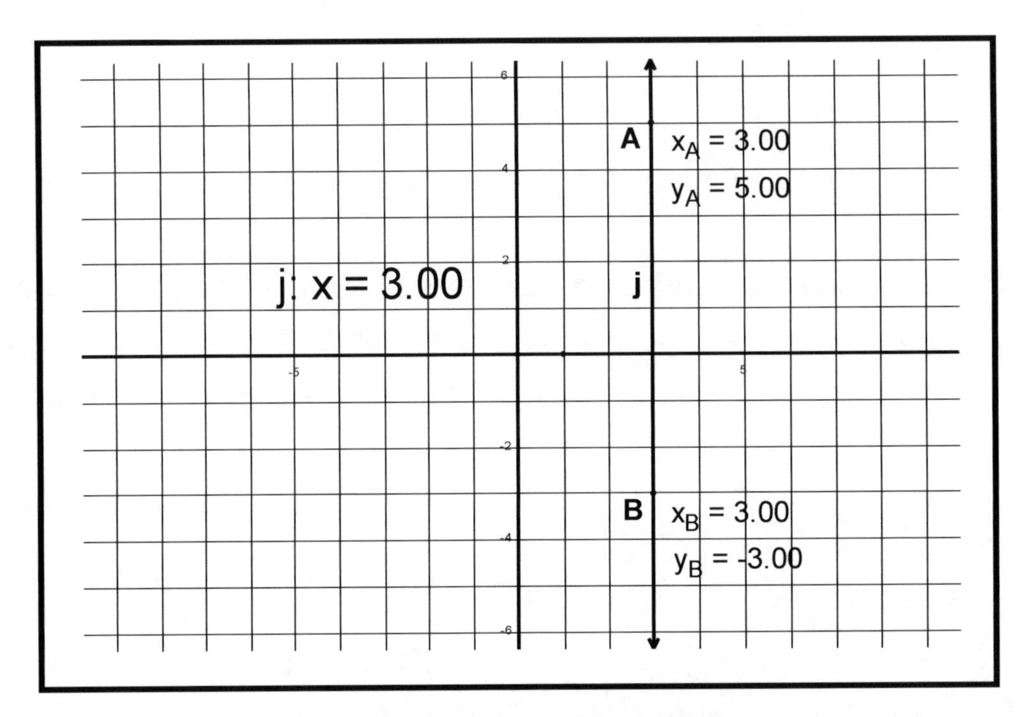

SUGGESTED EXERCISES

Type the solutions to the following problems in the blank region of your sketch.

1. Draw the line whose equation is x = -2. What is the value of the x-coordinate of every point on this line?
2. Draw the line whose equation is x = 4. What is the value of the x-coordinate of every point on this line?
3. Given the equation of the line x = 0, what is the value of the x-coordinate of every point on this line? Which axis (y-axis or x-axis) does the equation x = 0 represent?

LAB #71: THE EQUATION OF A HORIZONTAL LINE ON THE COORDINATE PLANE

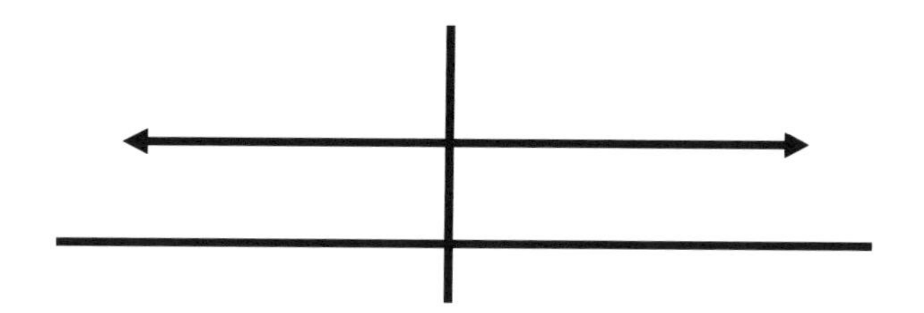

1. Select: [EDIT], [PREFERENCES], [TEXT], check the box [FOR ALL NEW POINTS] and click [OK].
2. Select: [GRAPH], [SHOW GRID], [GRAPH] and [SNAP POINTS].
3. Select: the Point tool and plot a point anywhere on the coordinate plane.
4. Select: the Arrow tool and click in the blank region to deselect.
5. Select: point A, the y-axis, [CONSTRUCT], [PERPENDICULAR LINE], [CONSTRUCT], [POINT ON PERPENDICULAR LINE] and deselect.
6. Click on point B, and keeping the left button depressed, drag it about two inches away from point A and deselect.
7. Select: points A, B, [MEASURE], [ABSCISSAE (x)] and deselect.
8. Select: points A, B, [MEASURE], [ORDINATES (y)] and deselect.
9. Click on point B, and keeping the left button depressed, drag it along line $\overleftrightarrow{AB}$. Observe the values that remain the same.
10. Select: the Text tool, double click in the blank region to open a dialog box and describe your observations about the values of the y-coordinates of points A and B.
11. Select: the Arrow tool and click in the blank region to deselect.
12. Select: line $\overleftrightarrow{AB}$, [MEASURE] and [EQUATION]. The horizontal line $\overleftrightarrow{AB}$ will be labeled as j and the equation of the horizontal line j will appear on your sketch.

13. Select: the Text tool, double click in the blank region to open a dialog box and (1) write the equation of the horizontal line j, (2) explain why the equation of the horizontal line is the same as the value of the y-coordinates of the points on this line.
14. Select: the Arrow tool, click on point A, and keeping the left button depressed, drag it up or down. Observe the values that remain equal as you do so.
15. Select: the Text tool, double click in the blank region to open a dialog box and explain how the equation of the horizontal line is related to the y-coordinate of points A or B.

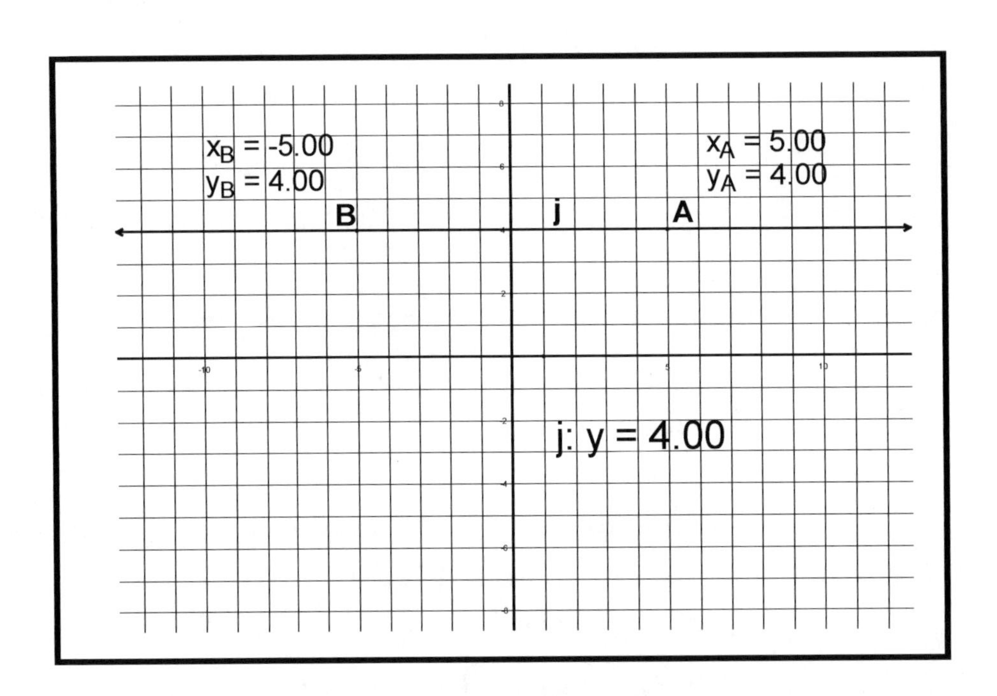

SUGGESTED EXERCISES

Type the solutions to the following problems in the blank region of your sketch.

1. Draw the line whose equation is y = 4. What is the value of the y-coordinate of every point on this line?
2. Draw the line whose equation is y = -2. What is the value of the y-coordinate of every point on this line?
3. Given the equation of the line y = 0, what is the value of the y-coordinate of every point on this line? Which axis (y-axis or x-axis) does equation y = 0 represent?

LAB #72: THE COORDINATES OF THE MIDPOINT OF A LINE SEGMENT

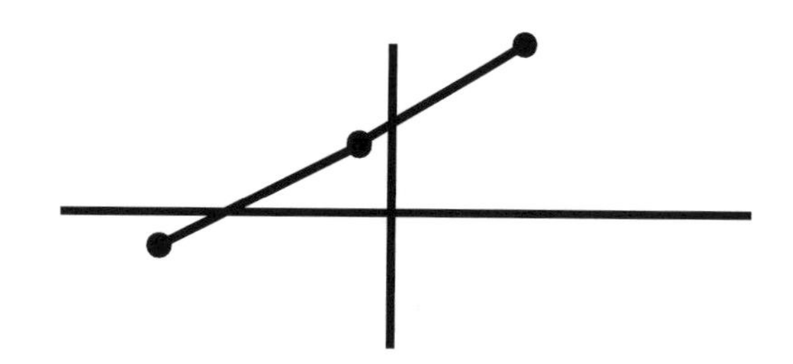

1. Select: [EDIT], [PREFERENCES], [TEXT], check the box [FOR ALL NEW POINTS] and click [OK].
2. Select: [GRAPH], [SHOW GRID], [GRAPH] and [SNAP POINTS].
3. Select: the Point tool and plot two points randomly.
4. Select: the Arrow tool and click in the blank region to deselect.
5. Select: points A, B, [CONSTRUCT], [SEGMENT], [CONSTRUCT], [MIDPOINT] and deselect.
6. Select: points A, B, midpoint C, [MEASURE], [ABSCISSAE (x)], and deselect.
7. Select: [NUMBER], [CALCULATE], [( ], click on the caption that shows the value of x_A, [+], click on the caption that shows the value of x_B, [ )], [÷], [2] and [OK].
8. Select: the Text tool, double click in the blank region to open a dialog box and describe your observations about the measure of the x-coordinate of the midpoint, x_C, and the value of $\frac{x_A + x_B}{2}$.
9. Select: the Arrow tool and click in the blank region to deselect.
10. Select: points A, B, midpoint C, [MEASURE], [ORDINATES (y)] and deselect.
11. Select: [NUMBER], [CALCULATE], [( ], click on the caption that shows the value of y_A, [+], click on the caption that shows the value of y_B, [ )], [÷], [2] and [OK].

12. Select: the Text tool, double click in the blank region to open a dialog box and describe your observations about the measure of the y-coordinate of the midpoint, y_C, and the value of $\frac{y_A + y_B}{2}$.
13. Select: the Arrow tool, click on point B, and, keeping the left button depressed, drag it. Do the same to drag point A. Observe the values that remain equal as you do so.
14. Select: the Text tool, double click in the blank region to open a dialog box and explain (1) what you have to add and divide by to find the x-coordinate of a midpoint, (2) what you have to add and divide by to find the y-coordinate of a midpoint.

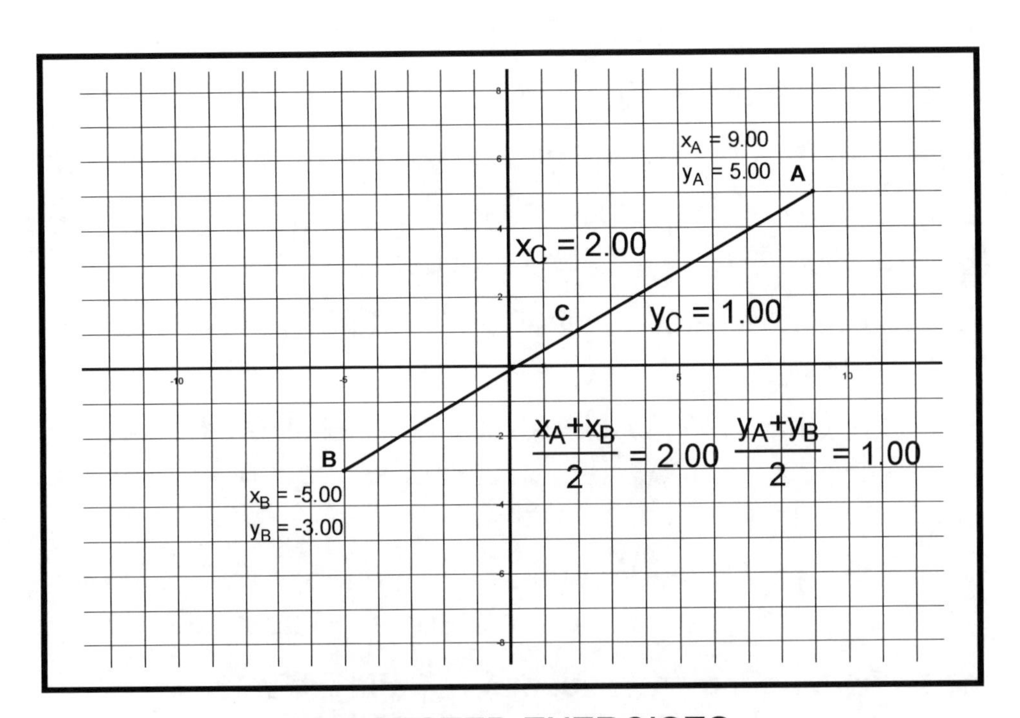

SUGGESTED EXERCISES

Type the solutions to the following problems in the blank region of your sketch.

1. Find the coordinates of the midpoint of a line segment $\overline{AB}$ whose endpoints are A (1, 5) and B (3, 9).
2. Find the coordinates of the center of a circle P if A (3, 4) and B (7, 8) are the endpoints of a diameter of a circle.
3. Midpoint M of a line segment $\overline{AB}$ has coordinates (6, 5). If the coordinates of point A are (4, 1), find the coordinates of point B.

LAB #73: THE DISTANCE BETWEEN THE ENDPOINTS OF A HORIZONTAL LINE SEGMENT

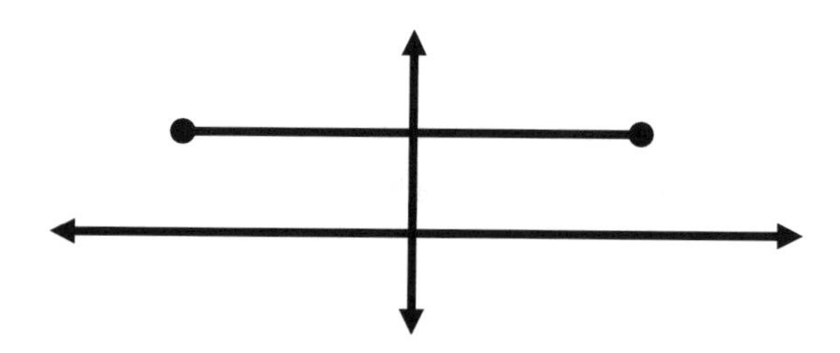

1. Select: [EDIT], [PREFERENCES], [TEXT], check the box [FOR ALL NEW POINTS] and click [OK].
2. Select: [GRAPH], [SHOW GRID], [GRAPH] and [SNAP POINTS].
3. Select: the Point tool and plot a point anywhere in the plane.
4. Select: the Arrow tool and click in the blank region to deselect.
5. Select: point A, the x-axis, [CONSTRUCT], [PARALLEL LINE], [CONSTRUCT], [POINT ON PARALLEL LINE] and deselect.
6. Click on point B, and keeping the left button depressed, drag it at least five units away from point A and deselect.
7. Select: line $\overleftrightarrow{AB}$, [DISPLAY] and [HIDE PARALLEL LINE].
8. Select: points A, B, [CONSTRUCT], [SEGMENT], [MEASURE], [LENGTH] and deselect.
9. Select: points A, B, [MEASURE], [ABSCISSAE (x)] and deselect.
10. Select: points A, B, [MEASURE], [ORDINATES (y)] and deselect.
11. Select: [NUMBER], [CALCULATE], [FUNCTIONS], [ABS], click on the caption that shows the value of x_B, [–], click on the caption that shows the value of x_A, [OK] and deselect.
12. Select: [NUMBER], [CALCULATE], [FUNCTIONS], [ABS], click on the caption that shows the value of x_A, [–], click on the caption that shows the value of x_B, [OK] and deselect.
13. Select: the Text tool, double click in the blank region to open a dialog box and (1) describe your observations about the measure of line segment $\overline{AB}$ and the values of $|x_A - x_B|$ and $|x_B - x_A|$, (2) explain why $|x_A - x_B|$ and $|x_B - x_A|$ are producing the same value.

14. Select: the Arrow tool, click on point A, and keeping the left button depressed, drag it. Do the same to drag point B. Observe the values that remain equal as you do so.

15. Select: the Text tool, double click in the blank region to open a dialog box and explain (1) which coordinates you have to subtract to find the distance between the endpoints of the horizontal line segment $\overline{AB}$, (2) why we have to use the absolute value symbols to represent distance on the coordinate plane.

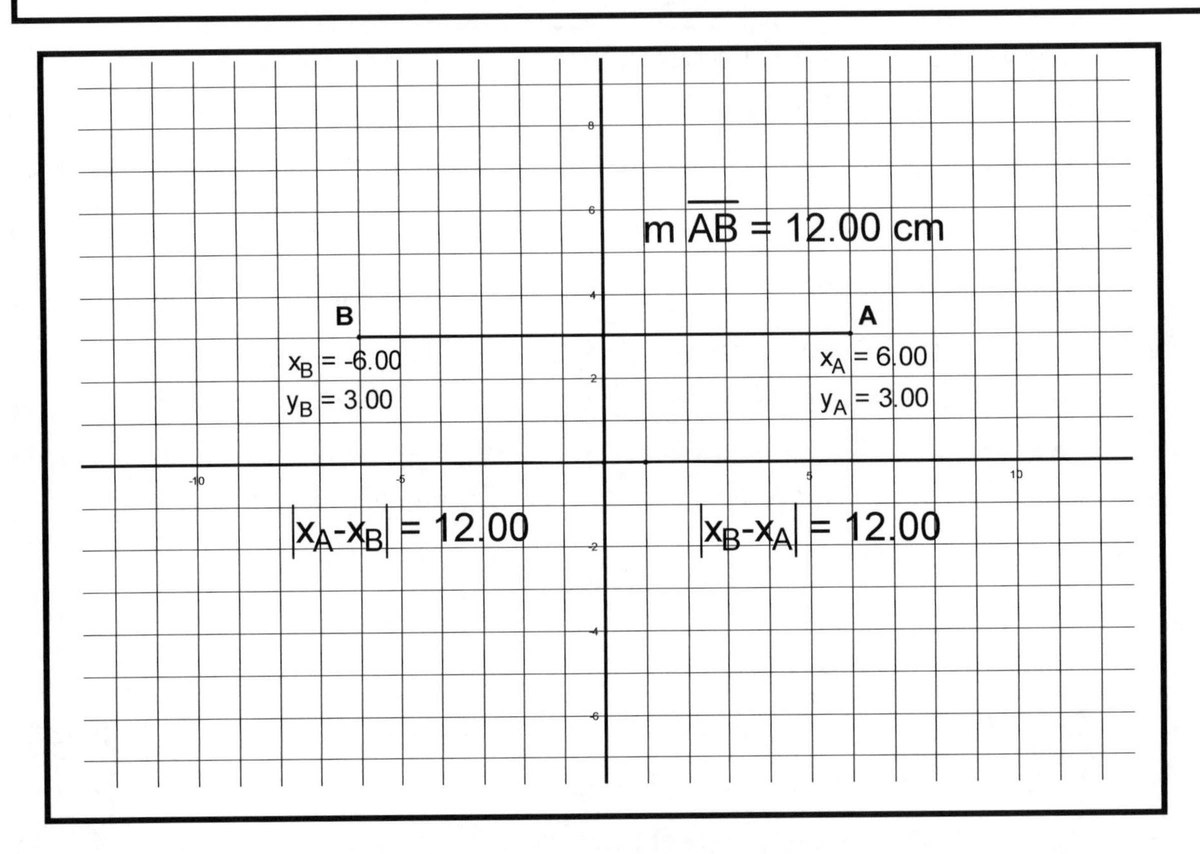

SUGGESTED EXERCISES

Type the solutions to the following problems in the blank region of your sketch.

1. Find the length of line segment $\overline{AB}$ joining the points A (2, -1) and C (7, -1).

2. Find the length of the line segment $\overline{BD}$ joining the points B (-2, 4) and D(5, 4).

3. A (4, 3) and B (x, 3) are the endpoints of the line segment $\overline{AB}$. If AB = 6, find the x-coordinate of the point B.

LAB #74: THE DISTANCE BETWEEN THE ENDPOINTS OF A VERTICAL LINE SEGMENT

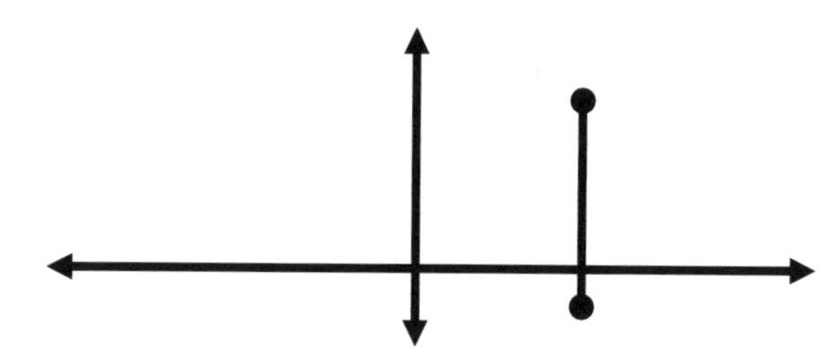

1. Select: [EDIT], [PREFERENCES], [TEXT], check the box [FOR ALL NEW POINTS] and click [OK].
2. Select: [GRAPH], [SHOW GRID], [GRAPH] and [SNAP POINTS].
3. Select: the Point tool and plot a point anywhere in the plane.
4. Select: the Arrow tool and click in the blank region to deselect.
5. Select: point A, the y-axis, [CONSTRUCT], [PARALLEL LINE], [CONSTRUCT] and [POINT ON PARALLEL LINE].
6. Click on point B, and keeping the left button depressed, drag it at least five units away from point A and deselect.
7. Select: line $\overleftrightarrow{AB}$, [DISPLAY] and [HIDE PARALLEL LINE].
8. Select: points A, B, [CONSTRUCT], [SEGMENT], [MEASURE], [LENGTH] and deselect.
9. Select: points A, B, [MEASURE], [ABSCISSAE (x)], and deselect.
10. Select: points A, B, [MEASURE], [ORDINATES (y)] and deselect.
11. Select: [NUMBER], [CALCULATE], [FUNCTIONS], [ABS], click on the caption that shows the value of y_B, [–], click on the caption that shows the value of y_A, [OK] and deselect.
12. Select: [NUMBER], [CALCULATE], [FUNCTIONS], [ABS], click on the caption that shows the value of y_A, [–], click on the caption that shows the value of y_B, [OK] and deselect.
13. Select: the Text tool, double click in the blank region to open a dialog box and (1) describe your observations about the measure of line segment $\overline{AB}$ and the values of $|y_A - y_B|$ and $|y_B - y_A|$, (2) explain why $|y_A - y_B|$ and $|y_B - y_A|$ produce the same value.

14. Select: the Arrow tool, click on point B, and keeping the left button depressed, drag it. Do the same to drag point A. Observe the values that remain equal as you so.
15. Select: the Text tool, double click in the blank region to open a dialog box and explain (1) which coordinates you have to subtract to find the distance between the endpoints of the vertical line segment $\overline{AB}$, (2) why we have to use the absolute value symbols to represent the distance on the coordinate plane.

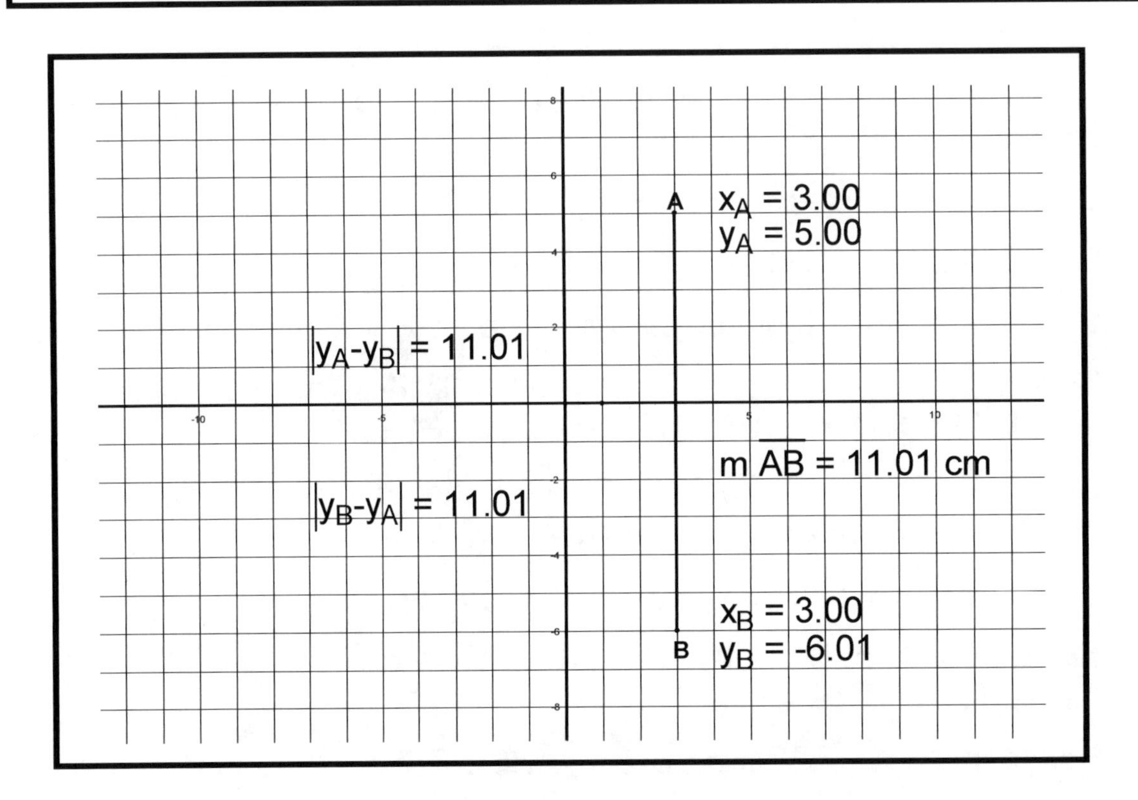

SUGGESTED EXERCISES

Type the solutions to the following problems in the blank region of your sketch.

1. Find the length of line segment $\overline{AB}$ joining the points A (1, 2) and C (1, 7).
2. Find the length of the line segment $\overline{BD}$ joining the points B (2, -3) and D (2, 4).
3. A (5, 2) and B (5, y) are the endpoints of the line segment $\overline{AB}$. If AB = 9, find the y-coordinate of the point B.

LAB #75: THE DISTANCE BETWEEN THE ENDPOINTS OF ANY LINE SEGMENT

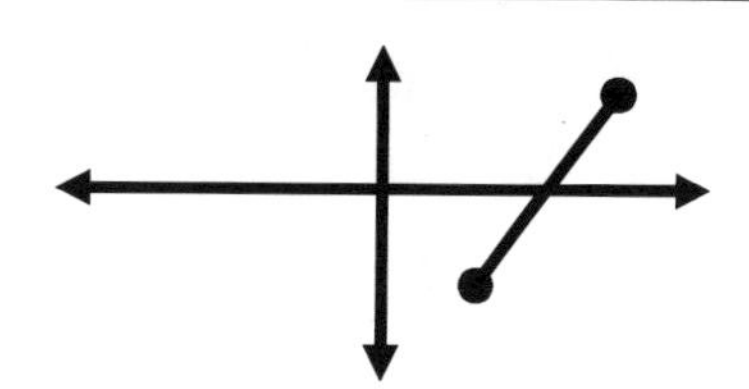

1. Select: [EDIT], [PREFERENCES], [TEXT], check the box [FOR ALL NEW POINTS] and click [OK].
2. Select: [GRAPH], [SHOW GRID], [GRAPH] and [SNAP POINTS].
3. Select: the Point tool and plot two points anywhere in the plane.
4. Select: the Arrow tool and click in the blank region to deselect.
5. Select: points A, B, [CONSTRUCT], [SEGMENT], [MEASURE], [LENGTH] and deselect.
6. Select: points A, B, [MEASURE], [ABSCISSAE (x)] and deselect.
7. Select: points A, B, [MEASURE], [ORDINATES (y)] and deselect.
8. Select: [NUMBER], [CALCULATE], [FUNCTIONS], [ABS], click on the caption that shows the value of x_B, [–], click on the caption that shows the value of x_A, [OK] and deselect.
9. Select: [NUMBER], [CALCULATE], [FUNCTIONS], [ABS], click on the caption that shows the value of y_B, [–], click on the caption that shows the value of y_A, [OK] and deselect.
10. Select: [NUMBER], [CALCULATE], [FUNCTIONS], [SQRT], click on the caption that shows $|X_B - X_A|$, [∧], [2], [+], click on the caption that shows $|Y_B - Y_A|$, [∧], [2] and [OK].
11. Click on point B, and keeping the left button depressed, drag it. Do the same to drag point A. Observe the values that remain equal as you do so.
12. Select: the Text tool, double click in the blank region to open a dialog box and explain what you have to do with the x-coordinates and the y-coordinates of two points to find the distance between these two points on the coordinate plane.
13. Select: the Arrow tool and click in the blank region to deselect.
14. Select: point B, the x-axis, [CONSTRUCT], [PERPENDICULAR LINE] and deselect.
15. Select: point A, the y-axis, [CONSTRUCT], [PERPENDICULAR LINE] and deselect.
16. Select: the perpendicular line through A, the perpendicular line through B, [CONSTRUCT], [INTERSECTION] and deselect.

17. Select: lines $\overleftrightarrow{AC}$, $\overleftrightarrow{BC}$, [DISPLAY] and [HIDE PERPENDICULAR LINES].
18. Select: points A, C, B, [CONSTRUCT], [SEGMENTS] and deselect.
19. Select: sides $\overline{AC}$, $\overline{BC}$, [MEASURE], [LENGTHS] and deselect.
20. Select: [NUMBER], [CALCULATE], [FUNCTIONS], [SQRT], click on the caption that shows m$\overline{AC}$, [∧], [2], [+], click on the caption that shows m$\overline{CB}$, [∧], [2] and [OK].
21. Select: the Text tool, double click in the blank region to open a dialog box and explain (1) the relationship between side $\overline{AC}$ and $|x_B - x_A|$, (2) the relationship between side $\overline{CB}$ and $|y_B - y_A|$, (3) the relationship between the Pythagorean Theorem $a^2 + b^2 = c^2$ and the distance formula $(x_B - x_A)^2 + (y_B - y_A)^2 = d^2$

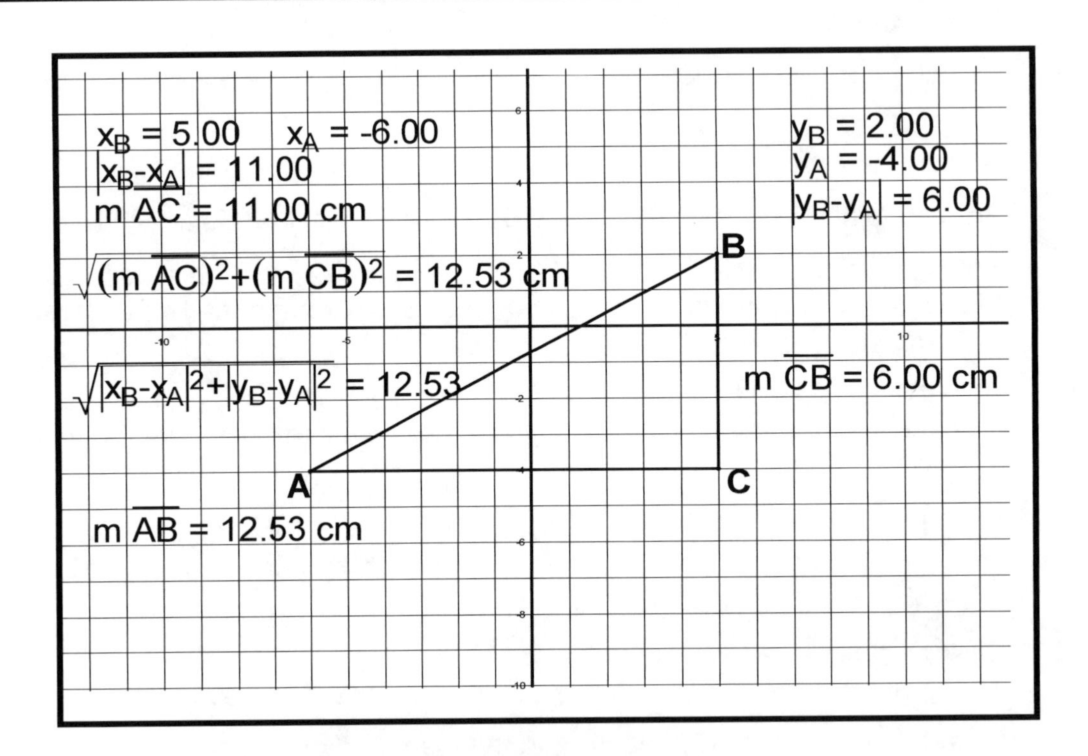

SUGGESTED EXERCISES

Type the solutions to the following problems in the blank region of your sketch.

1. Find the distance between the points A (5, -2) and B (9, 1).
2. Find the length of the radius of the circle to the nearest tenth whose center is C (1, 3) and that passes through the point D (7, -8).
3. The vertices of ΔABC are A (1, 1), B (9, 4), and C (1, 7). Show that Δ ABC is isosceles.

SOLUTIONS TO THE SUGGESTED EXERCISES

LAB #1: THE MIDSEGMENT OF A TRIANGLE

$m\overline{DE} = 3.74$ cm

$\frac{m\overline{CA}}{2} = 3.74$ cm

$m\overline{CA} = 7.48$ cm

1. $m\overline{DE} = \frac{m\overline{CA}}{2}$, thus $m\overline{DE} = 12/2 = 6$ in.
2. $m\overline{DE} = \frac{m\overline{CA}}{2}$, thus $2x = \frac{16}{2}$ or $2x = 8$ or $x = 4$.
3. $m\overline{DE} = \frac{m\overline{CA}}{2}$, thus $x + 20 = \frac{2(2x+10)}{2}$. Solving, we have $x + 20 = 2x + 10$. Combining like terms, we have $x = 10$. Substituting, $m\overline{DE} = 10 + 20 = 30$ and $m\overline{CA} = 2\{2(10) + 10\} = 2(30) = 60$

LAB #2:THE SUM OF THE ANGLES OF A TRIANGLE

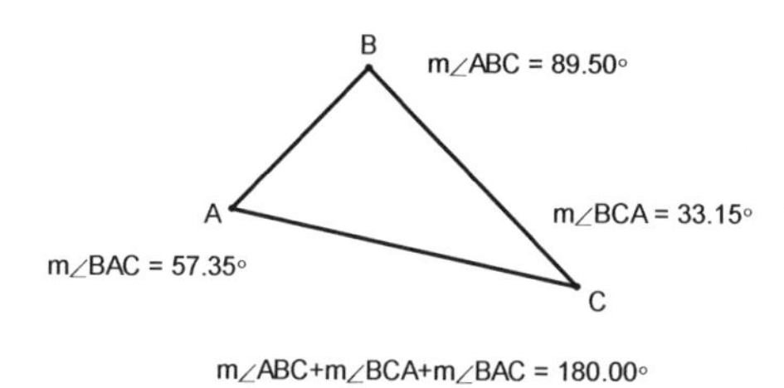

1. $m\angle C = 180 - (43 + 112) = 25°$.
2. Use the equation $2x + 98 + 52 = 180$. Solving, we have $2x = 180 - 98 - 52$. Simplifying, we have $2x = 30$, or $x = 15$.
3. Use the equation $(2x + 10) + (3x + 20) + 45 = 180$. Combining like terms, we have $5x + 75 = 180$, and then $5x = 105$, or $x = 21$. $m\angle A = 2x + 10 = 2(21) + 10 = 52°$ and $m\angle B = 3x + 20 = 3(21) + 20 = 83°$.

LAB #3: THE BISECTOR OF AN ANGLE

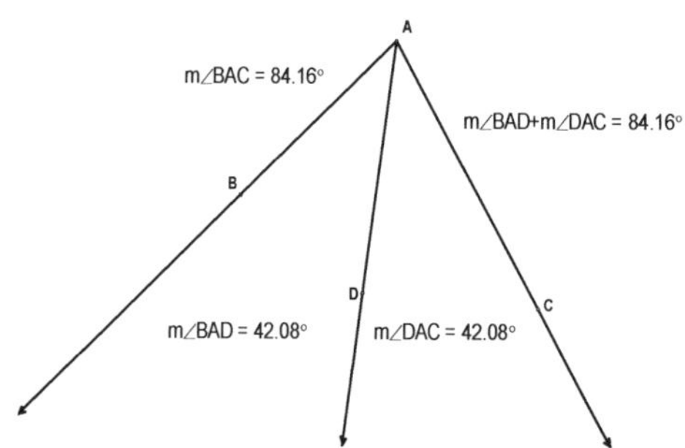

1. $m\angle BAD = 43°$.
2. Use the equation $3x - 4 = 38$. Solving, we have $3x = 42$ or $x = 14$.
3. Use the equation $2x + 10 = 4x$. Subtracting $2x$ from both sides, we have $10 = 2x$ or $x = 5$. Use this value of x to find $m\angle BAD = 4x = 4(5) = 20°$.
4. 40°.

LAB #4: THE MIDPOINT OF A LINE SEGMENT

$m\overline{AB} = 10.98\text{cm}$

A C B

$m\overline{AC} = 5.49\text{cm}$ $m\overline{CB} = 5.49\text{cm}$

$m\overline{AC} + m\overline{CB} = 10.98\text{cm}$

1. **$m\overline{AC} = 7$.**
2. **Use the equation $2x + 5 = 21$. Subtracting 5 from both sides, we have $2x = 16$, or $x = 8$.**
3. **Use the equation, $3x + 40 = x + 80$. Subtracting x from both sides, we have $2x + 40 = 80$. Subtracting 40 from both sides, we have $2x = 40$, or $x = 20$.**
   $m\overline{BC} = x + 80 = 20 + 80 = 100$. $m\overline{AB} = 3x + 40 = 3(20) + 40 = 100$.
4. **$4x + 6$.**

LAB #5: THE PERPENDICULAR BISECTOR OF A LINE SEGMENT

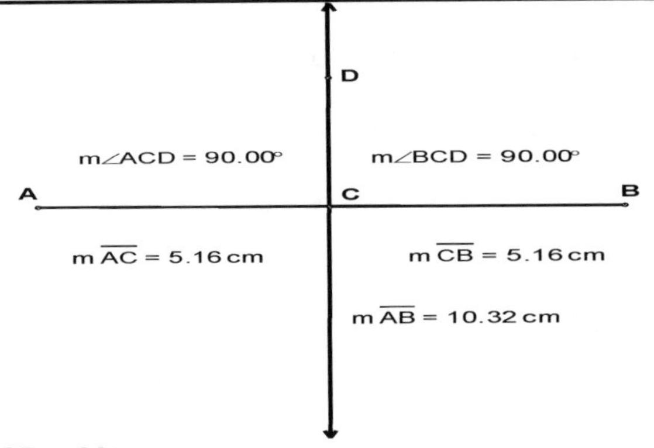

1. **$m\angle ACD = 90°$.**
2. **Use the equation $5x = 90$, or $x = 18$.**
3. **Since $\overline{AC} \cong \overline{BC}$, both segments are 3.22.**
4. **Since $\overline{AC} \cong \overline{BC}$, use the equation, $4x + 2 = 14$. Subtracting 2 from both sides, we have $4x = 12$, or $x = 3$.**

LAB #6: THE PERIMETER OF A POLYGON

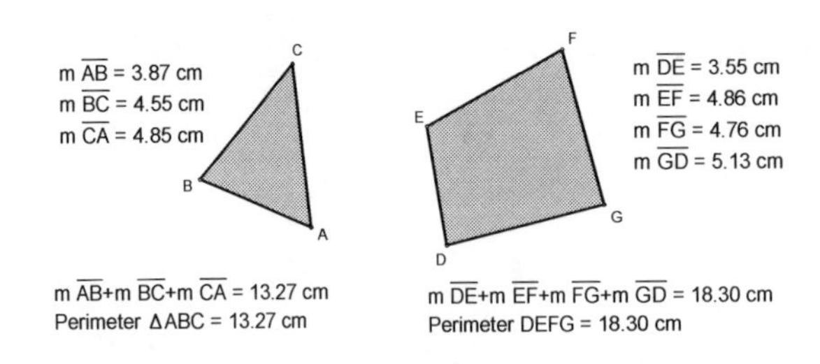

1. **Adding the three sides, we have 14.**
2. **Subtracting the two known sides from the perimeter, we have $17 - 8 - 5 = 4$. Also, $17 - (8 + 5) = 4$.**
3. **Use the equation $4 + 7 + 3 + 2x = 30$. Combining like terms, we have $14 + 2x = 30$. Subtracting 14 from both sides, we have $2x = 16$ or $x = 8$. $DG = 2x = 2(8) = 16$.**
4. **Use the equation $2x + 4 + 3x - 2 + x + 6 = 20$. Combining like terms we have $6x + 8 = 20$. Subtracting 8 from both sides, we have $6x = 12$ or $x = 2$. $m\overline{AB} = 2x + 4 = 2(2) + 4 = 8$. $m\overline{BC} = 3x - 2 = 3(2) - 2 = 4$. $m\overline{AC} = x + 6 = 2 + 6 = 8$.**

LAB #7: THE ALTITUDE OF A TRIANGLE

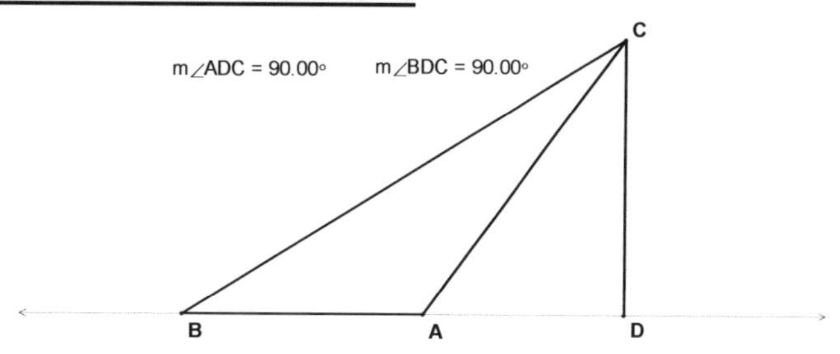

1. $m\angle ADC = 90°$.
2. Use the equation $2x = 90$ or $x = 45$.
3. Use the equation $4x - 10 = 90$. Adding 10 to both sides, we have $4x = 100$ or $x = 25$.

LAB #8: THE MEDIANS OF A TRIANGLE

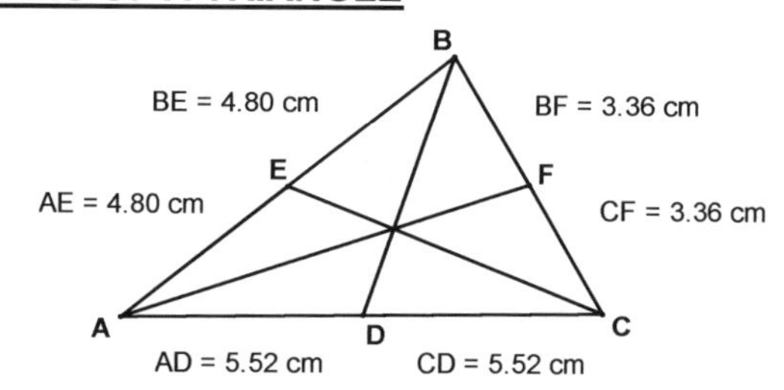

1. $m\overline{BE} = 3$.
2. Use the equation $3x = 21$ or $x = 7$.
3. Use the equation $4x + 20 = 2x + 30$. Subtracting 20 from both sides, we have $4x = 2x + 10$. Subtracting 2x from both sides, we have $2x = 10$ or $x = 5$. $m\overline{BF} = 4(5) + 20 = 40$. $m\overline{CF} = 2(5) + 30 = 40$. $m\overline{BC} = 40 + 40 = 80$.

LAB #9: COMPLEMENTARY ANGLES

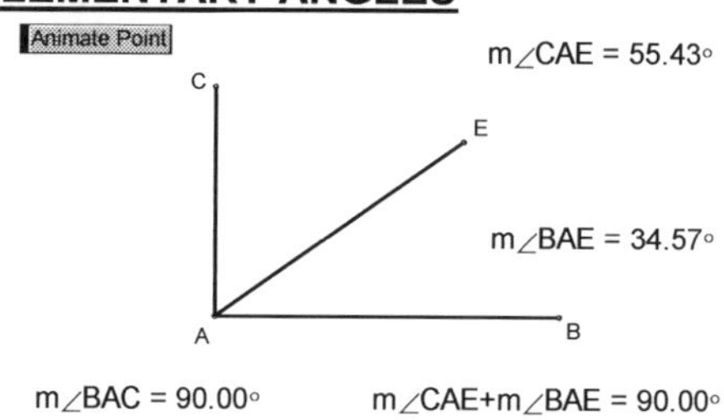

1. Subtract 37° from 90° to get 53°.
2. Use the equation $2x + 10 + x + 20 = 90$. Combining like terms, we have $3x + 30 = 90$. Subtracting 30 from both sides, we have $3x = 60$ or $x = 20$. $m\angle CAE = 2x + 10 = 2(20) + 10 = 50$. $m\angle BAE = x + 20 = 20 + 20 = 40$.
3. Since one angle is twice as big as the other, we may let x represent one angle and 2x represent the other. Use the equation $2x + x = 90$. Combining like terms, we have $3x = 90$ or $x = 30$. The angles are $x = 30°$ and $2x = 60°$. Both angles are needed to answer the question.

LAB #10: SUPPLEMENTARY ANGLES

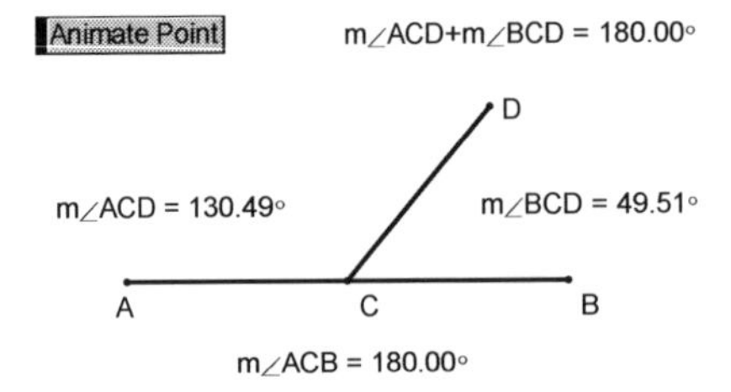

1. Subtract 69° from 180° to get 111°.
2. Use the equation $x + 3x = 180$. Combining like terms, we have $4x = 180$ or $x = 45$. $m\angle ACD = x = 45°$. $m\angle BCD = 3x = 3(45) = 135°$.
3. Use the equation $x + 30 + 3x - 10 = 180$. Combining like terms, we have $4x + 20 = 180$. Subtracting 20 from both sides, we have $4x = 160$ or $x = 40$. $m\angle A = x + 30 = 40 + 30 = 70$. $m\angle B = 3x - 10 = 3(40) - 10 = 110$.
4. Since one angle is 40° more than the other, we may let x represent one angle and $x + 40$ to represent the other. Use the equation $x + x + 40 = 180$. Combining like terms, we have $2x + 40 = 180$. Subtracting 40 from each side, we have $2x = 140$, or $x = 70$.

LAB #11: VERTICAL ANGLES

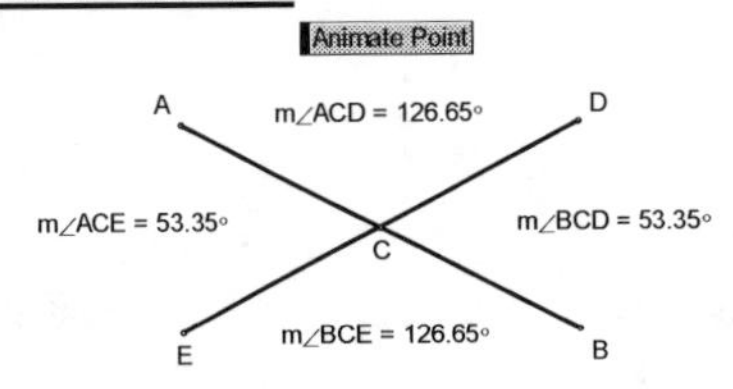

1. $m\angle BCD = 61°$.
2. Use the equation $5x = 3x + 10$. Subtracting 3x from both sides, we have $2x = 10$ or $x = 5$. Since $m\angle ACE$ is represented by 5x and $x = 5$, $m\angle ACE = 25°$.
3. Use the equation $7x + 16 = 3x + 48$. Subtracting 3x from both sides, we have $4x +16 = 48$. Subtracting 16 from both sides, we have $4x = 32$ or $x = 8$. Since $m\angle NTS$ is represented by $3x + 48$ and $x = 8$, $m\angle NTS = 72°$.

LAB #12: PARALLEL LINES – THE CORRESPONDING ANGLES

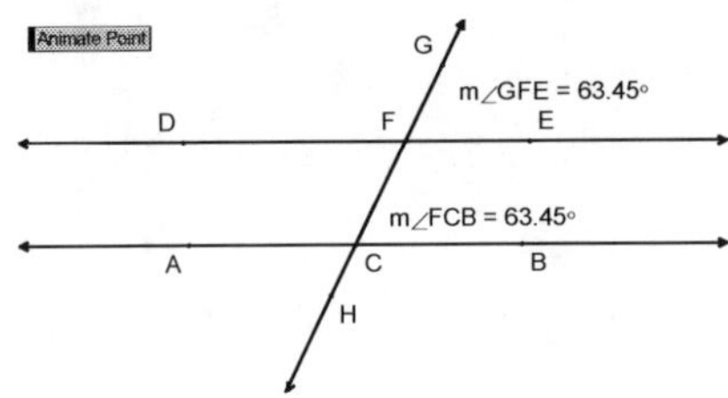

1. $m\angle FCB = 73°$.
2. Use the equation $4x = 120$, or $x = 30$.
3. Use the equation $2x + 40 = 3x + 20$. Subtracting 20 from both sides, we have $2x + 20 = 3x$. Subtracting 2x from both sides, we have $20 = x$. Since $m\angle GFE$ is represented by $2x + 40$, and $x = 20$, $m\angle GFE = 2(20) + 40 = 80$. Since $m\angle FCB = 3x + 20$ and $x = 20$, $m\angle FCB = 3(20) + 40 = 80$.

LAB #13: PARALLEL LINES – ALTERNATE INTERIOR ANGLES

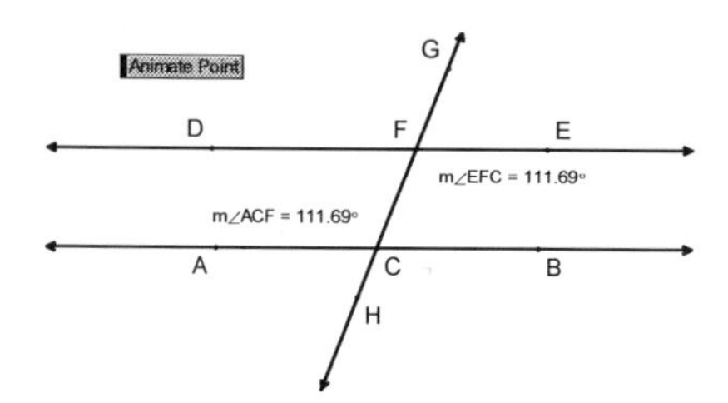

1. $m\angle ACF = 78°$.
2. Use the equation $3x = 135$ or $x = 45$.
3. Use the equation $7x = 3x + 60$. Subtracting $3x$ from both sides, we have $4x = 60$ or $x = 15$. Since $m\angle DFC = 7x$, and $x = 15$, $m\angle DFC = 105$. Since $m\angle BCF = 3x + 60$, $m\angle BCF = 3(15) + 60 = 105$.

LAB #14: PARALLEL LINES – THE INTERIOR ANGLES ON THE SAME SIDE OF THE TRANSVERSAL

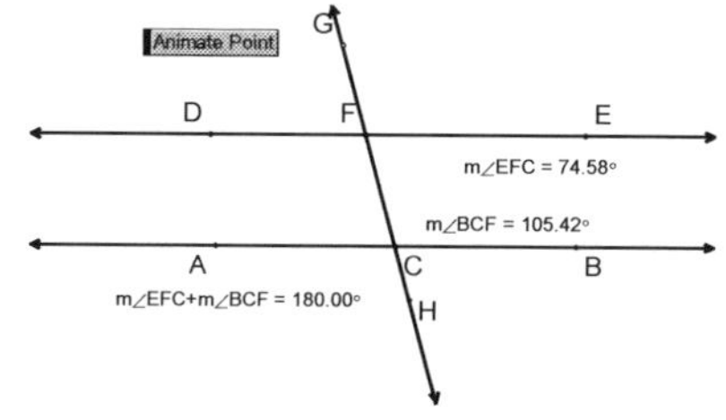

1. $m\angle BCF = 180 - 78 = 102°$.
2. Use the equation $3x + 114 = 180$. Subtracting 114 from both sides, we have $3x = 66$ or $x = 22$.
3. Use the equation $3x + 40 + 2x = 180$. Combining like terms, we have $5x = 140$, or $x = 28$. Since $m\angle ACF = 2x$, and $x = 28$, $m\angle ACF = 56°$. Since $m\angle DFC = 3x + 40$, and $x = 28$, we have $m\angle DFC = 124°$.

LAB #15: THE EXTERIOR ANGLE OF A TRIANGLE

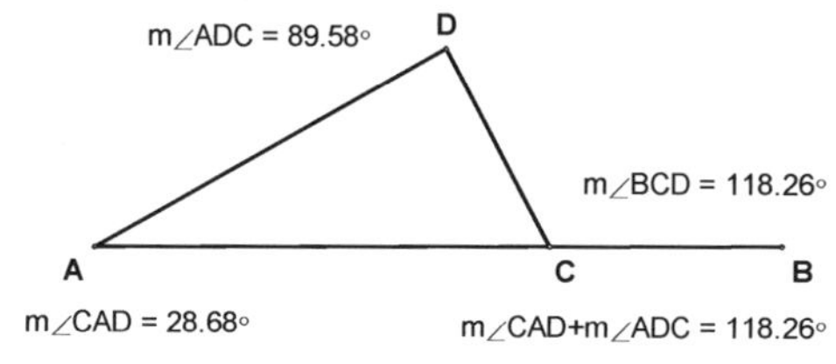

1. The exterior angle at vertex C is the sum of $m\angle ADC +\ m\angle DAC = 50 + 85 = 135°$.
2. Use the equation $90 + x = 150$. Solving for x, we have $x = 60$. So $m\angle DAC = 60°$.
3. Use the equation $2x + 10 = 140$. Subtracting 10 from each side, we have $2x = 130$ or $x = 65$. $m\angle A = x = 65°$. $m\angle D = x + 10 = 65 + 10 = 75°$.

LAB #16: THE SUM OF THE INTERIOR ANGLES OF A QUADRILATERAL

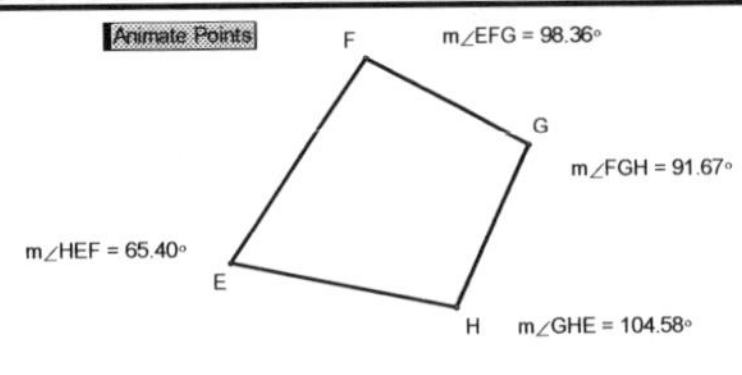

1. **$m\angle H = 360° - 73° - 64° - 142° = 81°$, or $360° - (73° + 64° + 142°) = 81°$.**
2. **Use the equation $81 + 70 + 139 + 2x = 360$. Combining like terms, we have $290 + 2x = 360$. Subtracting 290 from both sides, we have $2x = 70$ or $x = 35$.**
3. **Use the equation $46 + 93 + 2x + 6 + 3x - 5 = 360$. Combining like terms, we have $5x + 140 = 360$. Subtracting 140 from each side, we have $5x = 220$ or $x = 44$. $m\angle G = 2(44) + 6 = 94°$, and $m\angle H = 3(44) - 5 = 127°$.**

LAB #17: THE SUM OF THE INTERIOR ANGLES OF A POLYGON

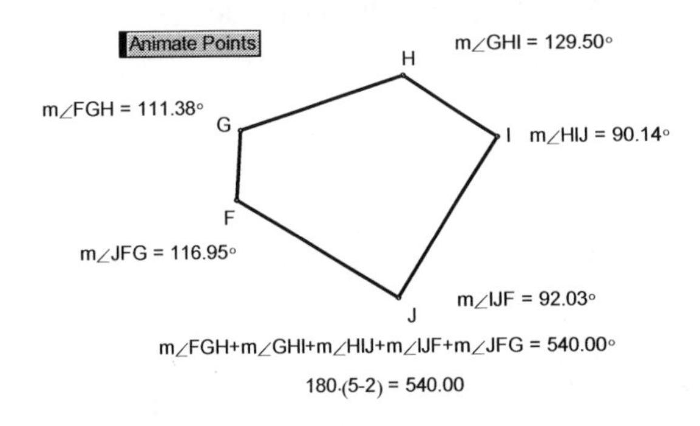

1. **$180(6 - 2) = 720°$.**
2. **$180(12 - 2) = 1800°$.**
3. **Use the equation $180(n - 2) = 2520$. Using the distributive property, we have $180n - 360 = 2520$. Adding 360 to each side, we have $180n = 2880$ or $n = 16$.**

LAB #18: THE SUM OF THE EXTERIOR ANGLES OF A TRIANGLE

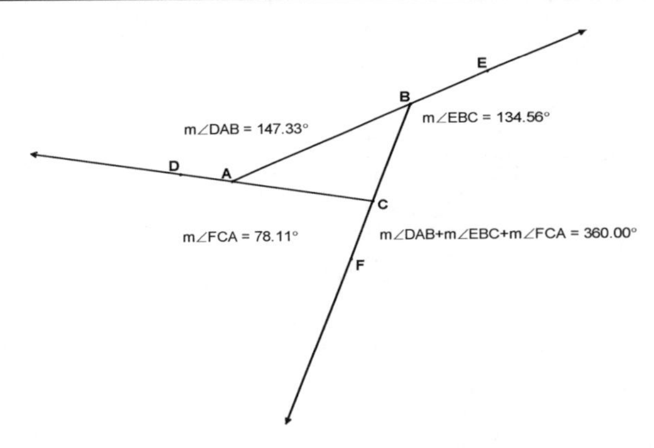

1. **$m\angle CBE = 360 - 112 - 84 = 164°$. Or $360 - (112 + 84) = 164°$.**
2. **Use the equation $8x + 120 + 80 = 360$. Combining like terms, we have $8x + 200 = 360$. Subtracting 200 from both sides, we have $8x = 160$ or $x = 20$. Using this value of x, we have $m\angle BAD = 8x = 8(20) = 160°$.**
3. **Use the equation $x + 10 + 3x - 20 + 2x + 40 = 360$. Combining like terms, we have $6x + 30 = 360$. Subtracting $6x = 330$ or $x = 55$. $m\angle 1 = x + 10 = 55 + 10 = 65$, $m\angle 2 = 3x - 20 = 3(55) - 20 = 145°$, and $m\angle 3 = 2x + 40 = 2(55) + 40 = 150°$.**

LAB #19: THE SUM OF THE EXTERIOR ANGLES OF A QUADRILATERAL

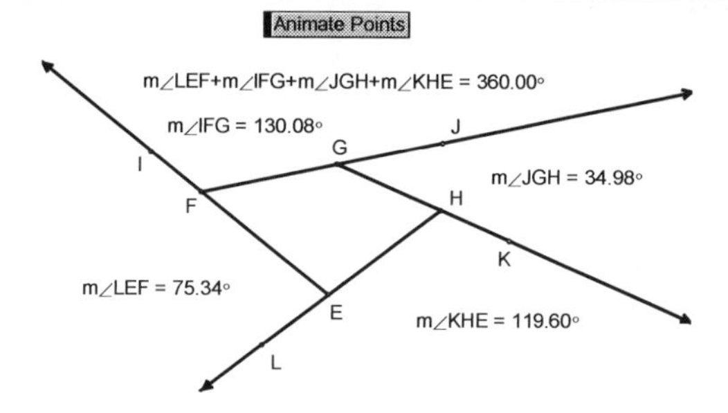

1. $m\angle 4 = 360 - (101 + 112 + 56) = 91°$. Or $360 - 101 - 112 - 56 = 91°$.
2. Use the equation $x + 2x + 4x + 5x = 360$. Combining like terms, we have $12x = 360$ or $x = 30$. $m\angle BAE = x = 30°$, $m\angle CBF = 2x = 2(30) = 60°$, $m\angle DCG = 4x = 4(30) = 120°$, and $m\angle ADH = 5x = 5(30) = 150°$.
3. Use the equation $x + 10 + 3x - 20 + 2x + 40 + 5x = 360$. Combining like terms, we have $11x + 30 = 360$. Subtracting 30 from both sides, we have $11x = 330$ or $x = 30$. $m\angle 1 = x + 10 = 30 + 10 = 40°$, $m\angle 2 = 3x - 20 = 3(30) - 20 = 70°$, $m\angle 3 = 2x + 40 = 2(30) + 40 = 100°$ and $m\angle 4 = 5x = 5(30) = 150°$.

LAB #20: THE SUM OF THE EXTERIOR ANGLES OF A POLYGON

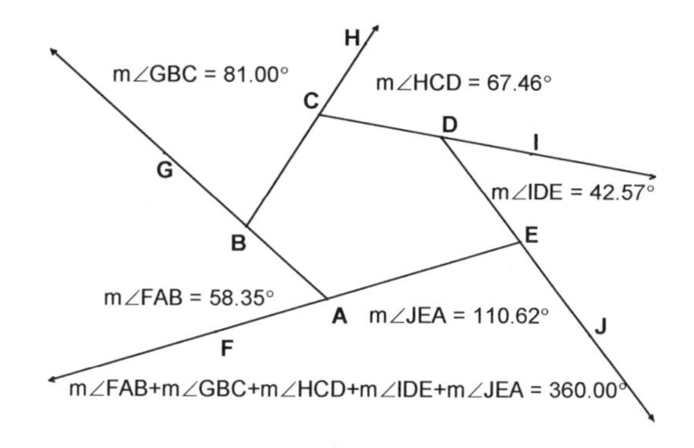

1. $m\angle 5 = 360 - (21 + 83 + 56 + 76) = 124°$.
2. Use the equation $x + 2x + 4x + 5x = 360$. Combining like terms, we have $12x = 360$ or $x = 30$. $m\angle 1 = x = 30°$, $m\angle 2 = 2x = 2(30) = 60°$, $m\angle 3 = 4x = 4(30) = 120°$, and $m\angle 4 = 5x = 5(30) = 150°$.
3. Use the equation $x + 10 + 3x - 20 + 2x + 40 = 360$. Combining like terms, we have $6x + 30 = 360$. Subtracting 30 from each side, we have $6x = 330$, or $x = 55$. $m\angle 1 = x + 10 = 55 + 10 = 65°$, $m\angle 2 = 3x - 20 = 3(55) - 20 = 145°$, and $m\angle 3 = 2x + 40 = 2(55) + 40 = 150°$.

LAB #21: THE PROPERTIES OF THE ISOSCELES TRIANGLE

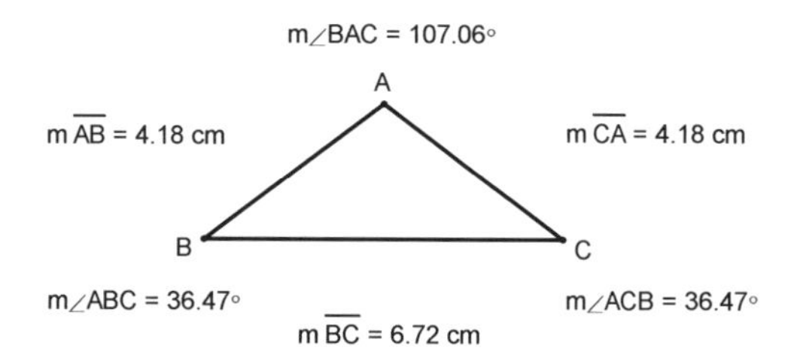

1. $m\overline{AC} = 7$.
2. Use the equation $x + 6 = 15$. Subtracting 6 from both sides, we have $x = 9$.

3. Use the equation $2x + 50 = 70$. Subtracting 50 from both sides, we have $2x = 20$ or $x = 10$. $m\angle C = 2x + 50 = 2(10) + 50 = 70°$.
4. Use the equation $3x + 15 = x + 50$. Subtracting x from both sides, we have $2x + 15 = 50$. Subtracting 15 from both sides, we have $2x = 35$, or $x = 17.5$. $m\angle ABC = 3x + 15 = 3(17.5) = 52.5 + 15 = 67.5°$, $\angle ACB = x + 50 = 17.5 + 50 = 67.5°$. The measure of the third angle can be found by computing $180 – 2(67.5) = 45°$.

LAB #22: THE PROPERTIES OF THE EQUILATERAL TRIANGLE

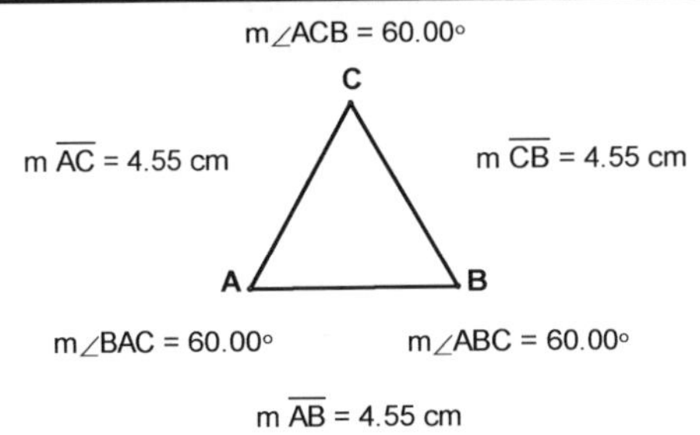

1. Each side is 6.
2. Use the equation $2x + 4 = 10$. Subtracting 4 from both sides, we have $2x = 6$ or $x = 3$.
3. Each angle is 60°. Use the equation $2x – 10 = 60$. Adding 10 to both sides, we have $2x = 70$ or $x = 35$.
4. Use the equation $4x – 14 = x + 22$. Subtracting x from each side, we have $3x – 14 = 22$. Adding 14 to both sides, we have $3x = 36$ or $x = 12$. $AC = 4x – 14 = 4(12) – 14 = 34$.

LAB #23: THE TRIANGLE INEQUALITY

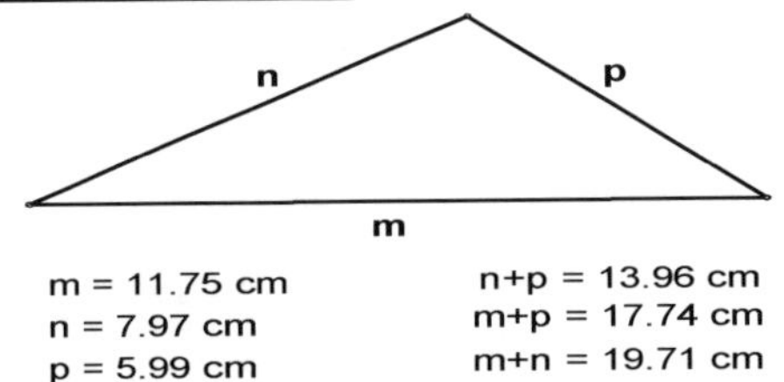

1. No, because the sum of the two shortest sides must be greater than the longest side of the triangle.
2. Choice 2 is the only possible set of sides where the sum of any two sides is greater than the third side.
3. The third side x, could one of the two shorter sides. In that case $x + 2 > 4$ or $x > 2$. If the third side x, is the largest side of the triangle, $2 + 4 > x$ or $6 > x$ or $x < 6$. The set of all possible values for the third side is any number greater than 2 but less than 6.

LAB # 24: THE PYTHAGOREAN THEOREM

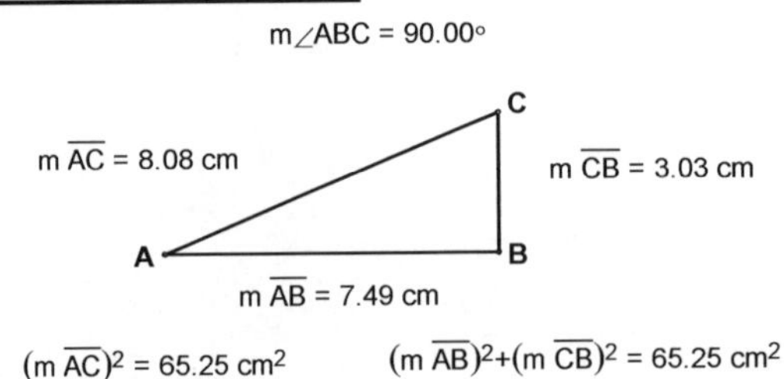

1. Use the equation, $3^2 + 4^2 = x^2$. So $x^2 = 9 + 16 = 25$.

Take the square root of both sides of the equation $x^2 = 25$ to get $x = 5$.

2. Use the equation $5^2 + x^2 = 13^2$. Squaring each term, we have $25 + x^2 = 169$. Subtracting 25 from each side, we have $x^2 = 144$. Taking the square root of each side, we have $x = 12$.
3. Use the equation $x^2 + (2x)^2 = 10^2$. Squaring each term, we have $x^2 + 4x^2 = 100$. Combining like terms, we have $5x^2 = 100$. Dividing both sides by 5, we have $x^2 = 20$. Taking the square root of each side, we have 4.5.

LAB #25: THE PROPERTIES OF REGULAR POLYGONS

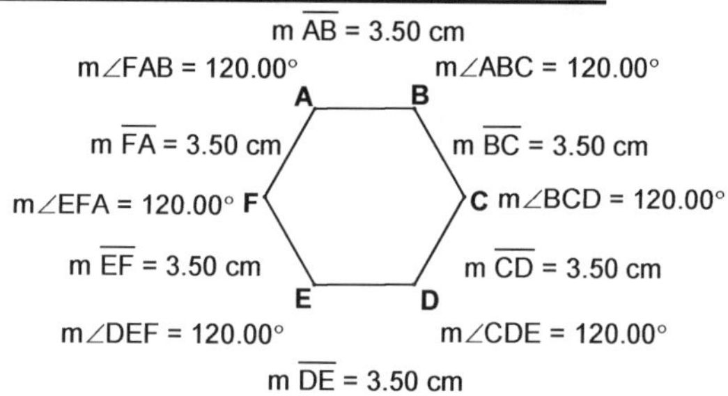

1. Choice 1. A regular must have all congruent sides and congruent angles.
2. A square and an equilateral triangle are regular polygons since they have all congruent sides and congruent angles.
3. Each of the other sides is also 13.2.

LAB #26: THE OPPOSITE ANGLES OF A PARALLELOGRAM

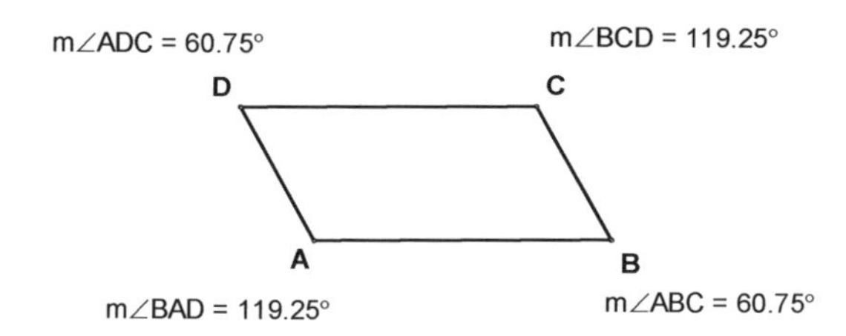

1. $m\angle C = 49°$.
2. Use the equation $3x = 123$. $x = 41$.
3. Use the equation $2x + 50 = 3x + 40$. Subtracting $2x$ from both sides, we have $50 = x + 40$. Subtracting 40 from both sides, we have $x = 10$. The $m\angle ABC = 2x + 50 = 2(10) + 50 = 70°$ and $m\angle ADC = 3x + 40 = 3(10) + 40 = 70°$.

LAB #27: THE CONSECUTIVE ANGLES OF A PARALLELOGRAM

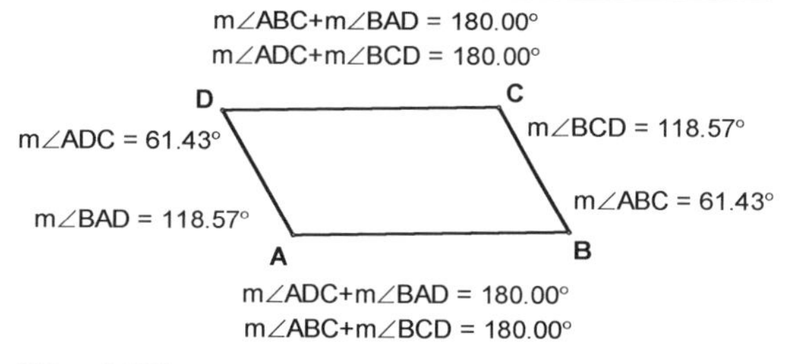

1. $m\angle B = 180 - 37 = 143°$.
2. Use the equation $50 + 4x = 180$. Subtracting 50 from both sides, we have $4x = 130$. Dividing by 4, we have 32.5
3. Use the equation $2x + 10 + 3x + 20 = 180$. Combining like terms, we have $5x + 30 = 180$. Subtracting 30 from both sides, we have $5x = 150$, or $x = 30$. $m\angle ABC = 3x + 20 = 3(30) + 20 = 110°$. The measure of the opposite angle $\angle CDA$ is 110°. The two adjacent angles,

$\angle BAD$ and $\angle BCD$ both contain 70° since they are both supplementary to the obtuse angles of the parallelogram which are 110°.

LAB # 28: THE OPPOSITE SIDES OF A PARALLELOGRAM

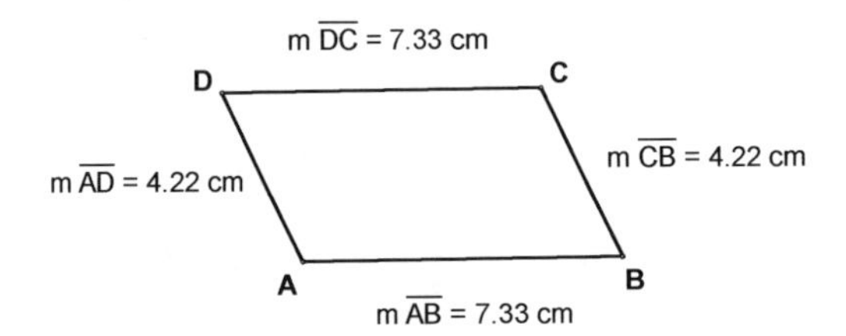

1. m$\overline{CD}$ = 12.
2. Use the equation 5x = 62. Dividing both sides by 5, we have x = 12.4.
3. Use the equation 3x + 8 = x + 42. Subtracting x from both sides, we have 2x + 8 = 42. Subtracting 8 from each side, we have 2x = 34 or x = 17.

LAB #29: THE DIAGONALS OF A PARALLELOGRAM

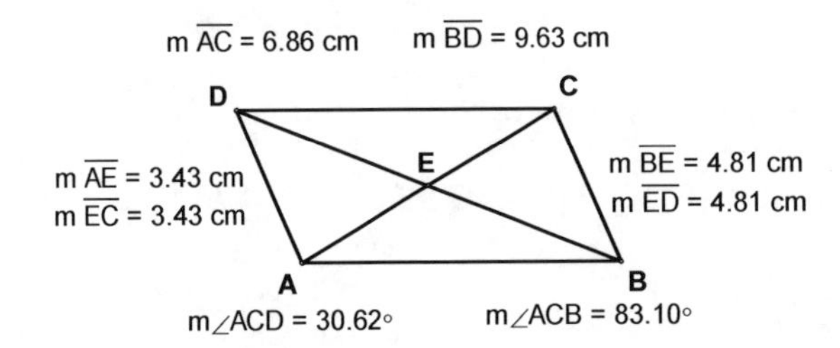

1. m$\overline{CE}$ = 4.
2. Use the equation 6x = 42. Dividing both sides by 6, we have x = 7.
3. Use the equation 5x = 3x + 7. Subtracting 3x from each side, we have 2x = 7 or x = 3.5. m$\overline{BE}$ = 5x = 5(3.5) = 17.5. m$\overline{DE}$ = 3x + 7 = 3(3.5) + 7 = 17.5.

LAB #30: THE PROPERTIES OF A RECTANGLE

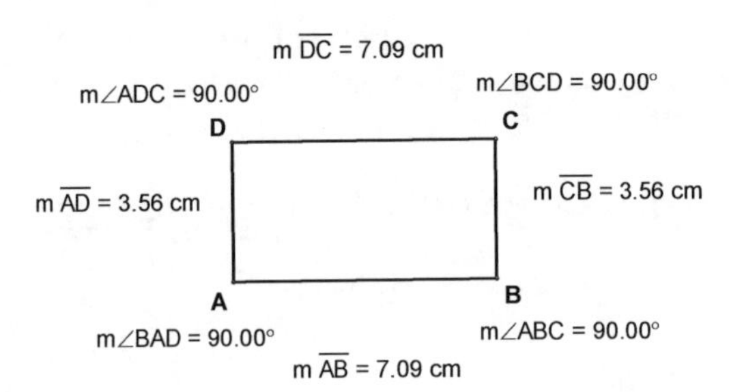

1. m$\overline{CD}$ = 7.
2. Use the equation 4x = 36. Dividing by 4, we have x = 9.
3. Use the equation 5x – 10 = 90. Adding 10 to both sides, we have 5x = 100 or x = 20.

LAB #31: THE DIAGONALS OF A RECTANGLE

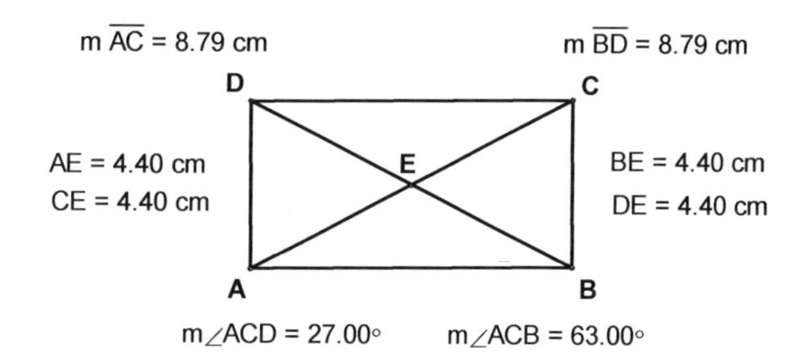

1. $m\overline{CE} = 9$.
2. Use the equation $4x + 6 + 64 = 90$. Combining like terms, we have $4x + 70 = 90$. Subtracting 70 from each side, we have $4x = 20$ or $x = 5$. $m\angle ACD = 4x + 6 = 4(5) + 6 = 26°$.
3. Use the equation $3x + 20 = x + 44$. Subtracting x from each side, we have $2x + 20 = 44$. Subtracting 20 from both sides, we have $2x = 24$ or $x = 12$. $m\overline{AC} = 3x + 20 = 3(12) + 20 = 56$. $m\overline{BD} = x + 44 = 12 + 44 = 56$.

LAB #32: THE PROPERTIES OF A SQUARE

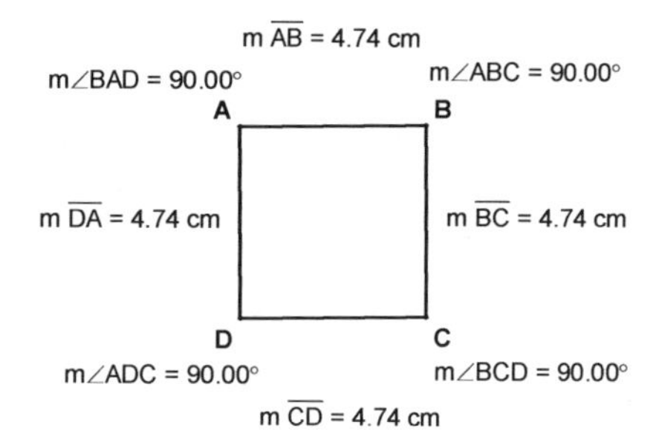

1. $m\overline{BC} = 5$.
2. Use the equation $4x = 29$. Dividing both sides by 4, we have $x = 7.25$.
3. Use the equation $8x - 6 = 5x + 12$. Subtracting 5x from each side, we have $3x - 6 = 12$. Adding 6 to each side, we have $3x = 18$ or $x = 6$. $m\overline{AD} = 8x - 6 = 8(6) - 6 = 48 - 6 = 42$. $m\overline{CD} = 5x + 12 = 5(6) + 12 = 42$. Each side of this square is 42.
4. Use the equation $3x + 30 = 90$. Subtracting 30 from each side, we have $3x = 60$ or $x = 20$.

LAB #33: THE DIAGONALS OF A SQUARE

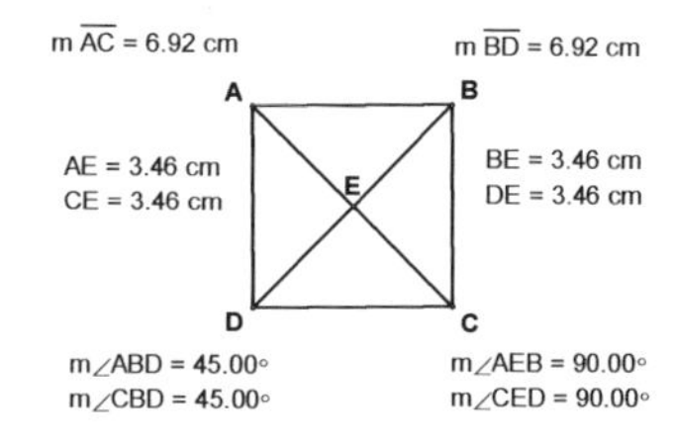

1. $m\overline{AE} = 5$.
2. Use the equation $3x = 45$ or $x = 15$.
3. Double the expression $3x + 4$ to get $6x + 8$.
4. Use the equation $2x = 45$ or $x = 22.5$.

LAB #34: THE PROPERTIES OF A RHOMBUS

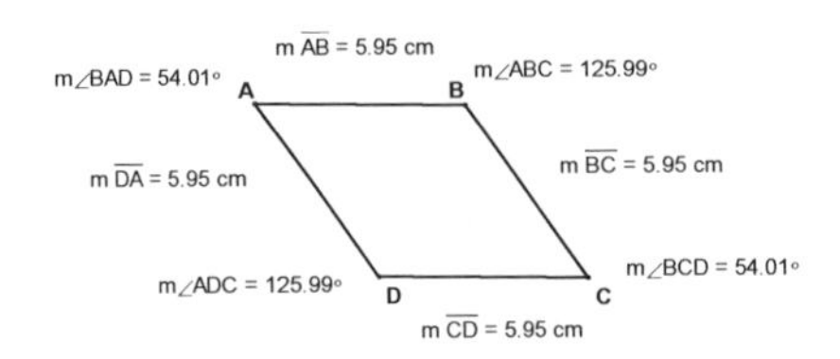

1. **m$\overline{BC}$ = 9.**
2. **Use the equation 3x = 48. Dividing both sides by 3, we have x = 16.**
3. **Use the equation 3x – 15 = x + 20. Subtracting x from both sides, we have 2x – 15 = 20. Adding 15 to both sides, we have 2x = 35 or x = 17.5.**
4. **Use the equation 3x = 2x + 50. Subtracting 2x from both sides, we have x = 50. m∠BAD = 3x = 3(50) = 150°. m∠BCD = 2x + 50 = 2(50) + 50 = 150°.**

LAB # 35: THE DIAGONALS OF A RHOMBUS

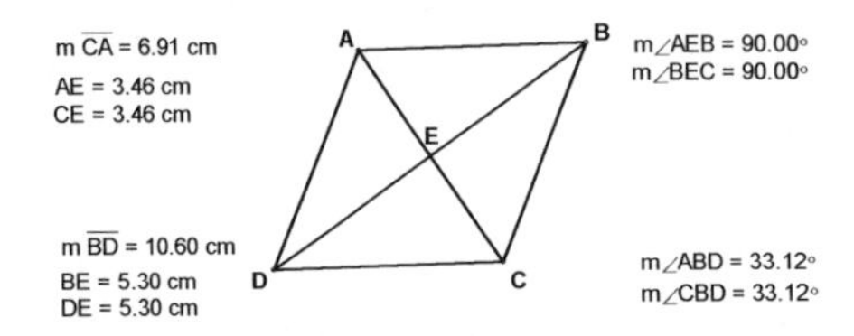

1. **m$\overline{CE}$ = 7.**
2. **Use the equation 4x = 52. Dividing both sides by 4, we have x = 13.**
3. **Use the equation 4x – 10 = x + 11. Subtracting x from both sides, we have 3x – 10 = 11. Adding 10 to both sides we have 3x = 21 or x = 7. m$\overline{BE}$ = 4x – 10 = 4(7) – 10 = 18. m$\overline{ED}$ = x + 11 = 7 = 11 = 18, so the entire diagonal $\overline{BD}$ is 36.**
4. **Use the equation 5x = 90 because the diagonals form a right angle. Dividing both sides by 5, we have x = 18.**

LAB #36: THE PROPERTIES OF A TRAPEZOID

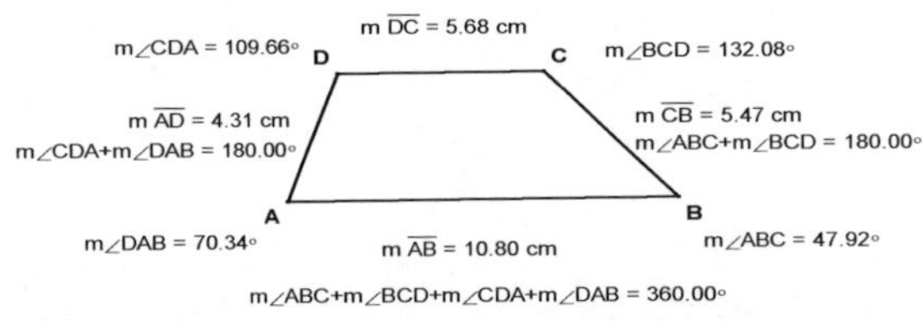

1. **m∠D = 180 – 51 = 129°. m∠C = 180 – 73 = 107°.**
2. **Use the equation x + 40 = 180. Subtracting 40 from each side, we have x = 140.**
3. **Use the equation x + 10 + 3x + 40 = 180. Combining like terms, we have 4x + 50 = 180. Subtracting 180 from each side, we have 4x = 130. Dividing both sides by 4, we have x = 32.5.**

LAB # 37: THE PROPERTIES OF THE ISOSCELES TRAPEZOID

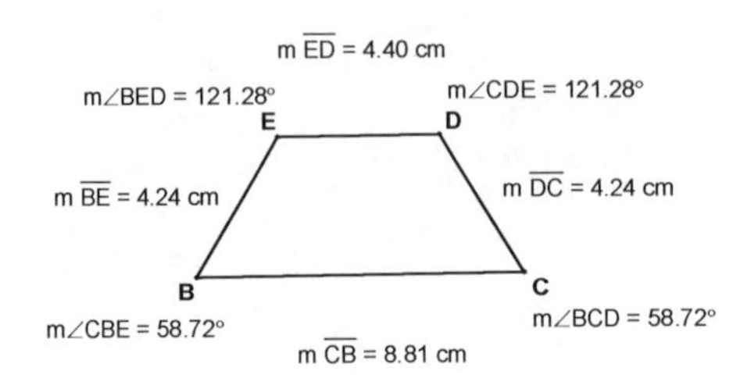

1. **m∠C = 48°.**
2. **Use the equation 5x = 120. Dividing both sides by 5, we have x = 24.**
3. **Use the equation x + 3 = 2x + 2. Subtracting x from both sides, we have 3 = x + 2. Subtracting 2 from both sides, we have x = 1. m$\overline{BE}$ = x + 3 = 1 + 3 = 4.**

LAB #38: THE DIAGONALS OF THE ISOSCELES TRAPEZOID

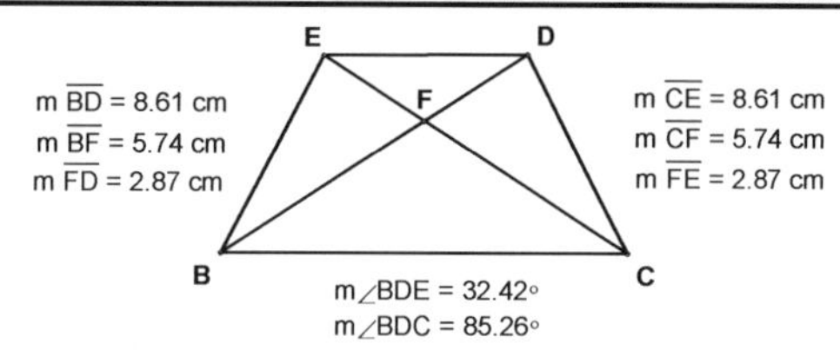

1. **m$\overline{BD}$ = 12.**
2. **Use the equation 3x = 21. Dividing both sides by 3, we have x = 7.**
3. **Use the equation 3x + 8 = x + 12. Subtracting x from each side, we have 2x + 8 = 12. Subtracting 8 from both sides, we have 2x = 4 or x = 2. m$\overline{CE}$ = 3x + 8 = 3(2) + 8 = 14.**

LAB #39: CORRESPONDING ANGLES OF SIMILAR TRIANGLES

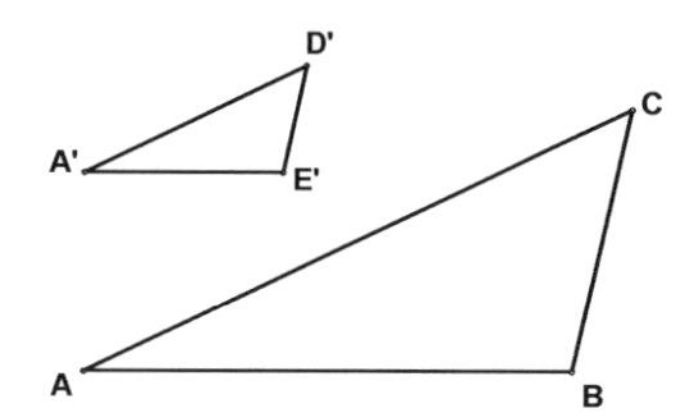

1. **m∠A'E'D' = 71° and m∠A'D'E' = 67°.**
2. **Use the equation 3x – 20 = 70. Adding 20 to both sides, we have 3x = 90 or x = 30.**
3. **Use the equation 2x + 20 = x + 35. Subtracting x from each side, we have x + 20 = 35. Subtracting 20 from each side, we have x = 15. m∠A'E'D' = 2x + 20 = 2(15) + 20 = 50°. m∠ABC = x + 35 = 15 + 35 = 50°.**

LAB #40: THE CORRESPONDING SIDES OF SIMILAR TRIANGLES

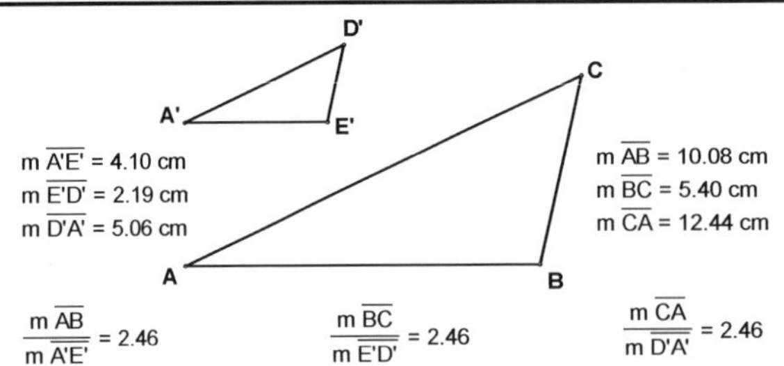

1. **Use the proportion $\frac{12}{8} = \frac{15}{A'E'}$. Cross multiplying, we have 12A'E' = 120. Dividing both sides by 12, we have A'E' = 10.**
2. **Use the proportion $\frac{8}{16} = \frac{2}{x}$. Cross multiplying, we have 8x = 32 or x = 4.**
3. **Use the proportion $\frac{12}{6} = \frac{24}{x}$. Cross multiplying, we have 12x = 144 or x = 12.**

LAB #41: CORRESPONDING PERIMETERS OF SIMILAR TRIANGLES

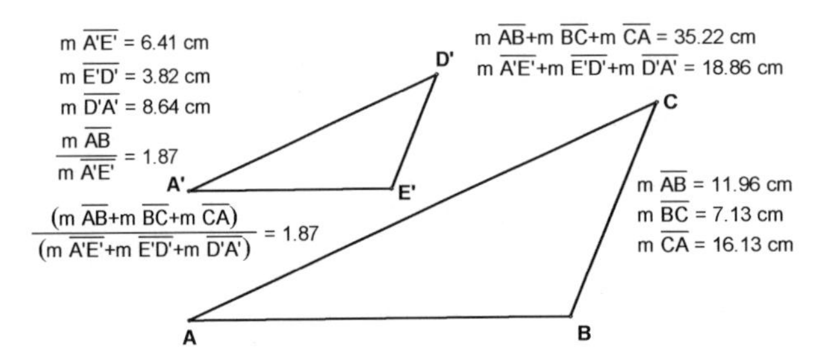

1. **$\frac{3}{4}$ or 3:4.**
2. **Let x = the smallest side of the similar triangle and use the equation $\frac{6}{x} = \frac{24}{12}$. Cross multiplying, we have 24x = 72 or x = 3.**
3. **Let x = the longest side of a similar triangle and use the equation $\frac{8}{x} = \frac{18}{54}$. Cross multiplying, we have 18x = 432. Dividing both sides of the equation by 18, we have x = 24.**

LAB #42: THE CORRESPONDING AREAS OF SIMILAR TRIANGLES

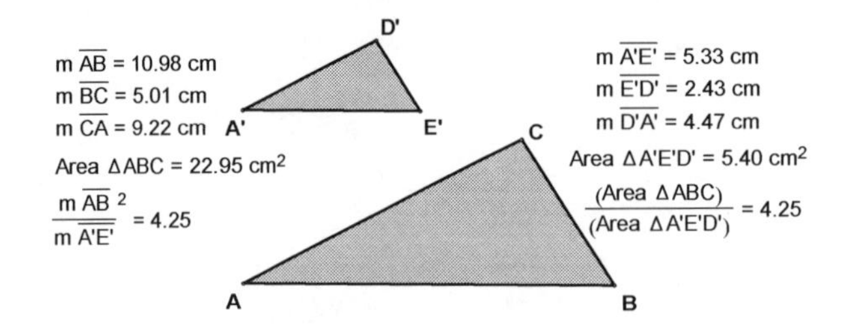

1. **$\frac{4}{9}$.**
2. **$\frac{5}{4}$.**

3. The ratio of the corresponding sides of these two similar triangles is $\frac{4}{6}$ or $\frac{2}{3}$. The ratio of the corresponding areas is $\frac{4}{9}$. Use the equation $\frac{4}{9} = \frac{20}{x}$. Cross multiplying, we have 4x = 180 or x = 45.

LAB #43: CONGRUENT TRIANGLES: S.S.S. ≅ S.S.S.

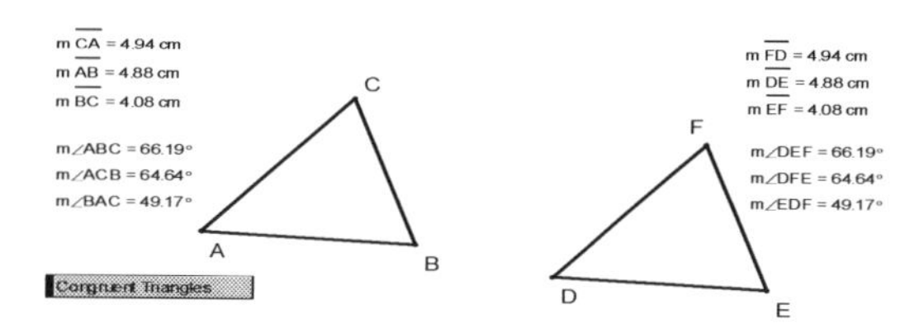

1. m$\overline{DE}$ = 7.
2. Use the equation 3x = 42. Dividing by 3, we have x = 14.
3. Use the equation 3x = 2x + 10. Subtracting 2x from each side, we have x = 10. m$\overline{AB}$ = 3x = 3(10) = 30. m$\overline{DE}$ = 2x + 10 = 2(10) + 10 = 30.
4. It is possible for two triangles to have three pairs of corresponding angles exactly the same and the triangles still not be congruent. The angles are what is controlling the shape. The triangles may be two different sizes.

LAB #44: CONGRUENT TRIANGLES: S.A.S. ≅ S.A.S.

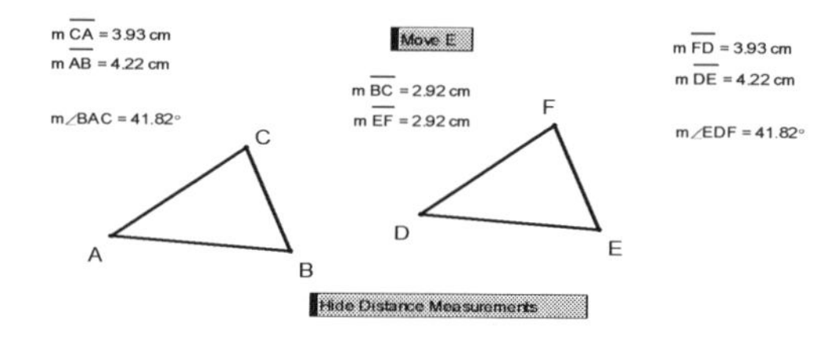

1. m$\overline{FD}$ = 4 cm, m∠EDF = 36° and m$\overline{DE}$ = 7 cm.
2. Use the equation 4x = 72. Dividing both sides of the equation by 4, we have x = 18.
3. Use the equation 2x + 5 = 3x + 4. Subtracting 2x from both sides, we have 5 = x + 4 or x = 1. m$\overline{AB}$ = 2x + 5 = 2(1) + 5 = 7. m$\overline{DE}$ = 3x + 4 = 3(1) + 4 = 7.
4. If the 60 degree is not in between the two given sides of 5 and 8, two different sized triangles may result.

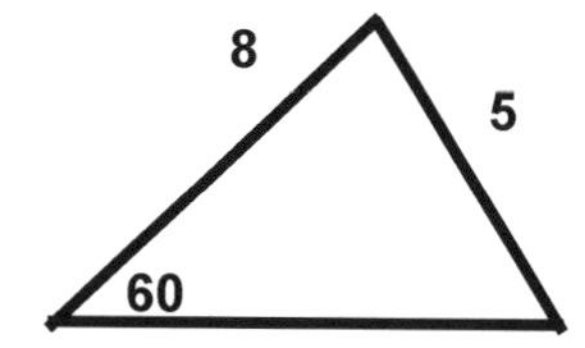

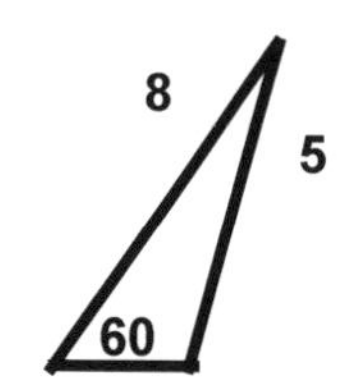

LAB #45: CONGRUENT TRIANGLES: A.S.A. ≅ A.S.A.

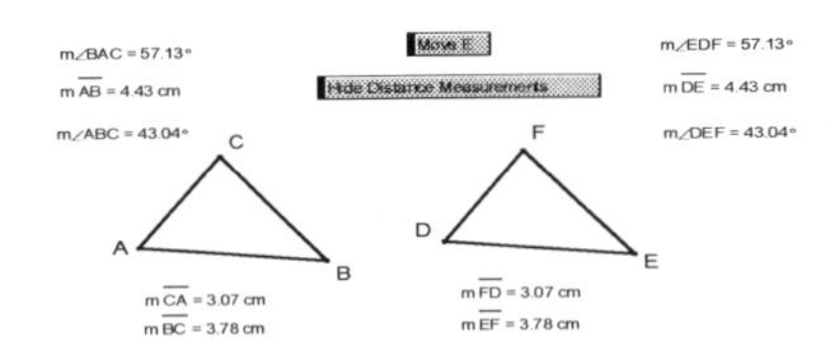

1. $m\angle EDF = 56°$, $m\overline{DE} = 5$ and $m\angle DEF = 61°$.
2. Use the equation $6x = 48$. $x = 8$.
3. Use the equation $8x - 6 = 5x + 12$. Subtracting 5x from each side, we have $3x - 6 = 12$. Adding 6 to both sides, we have $3x = 18$ or $x = 6$. $m\overline{AB} = 8x - 6 = 8(6) - 6 = 42$.
   $m\overline{DE} = 5x + 12 = 5(6) + 12 = 42$.

LAB #46: A REFLECTION IN A LINE

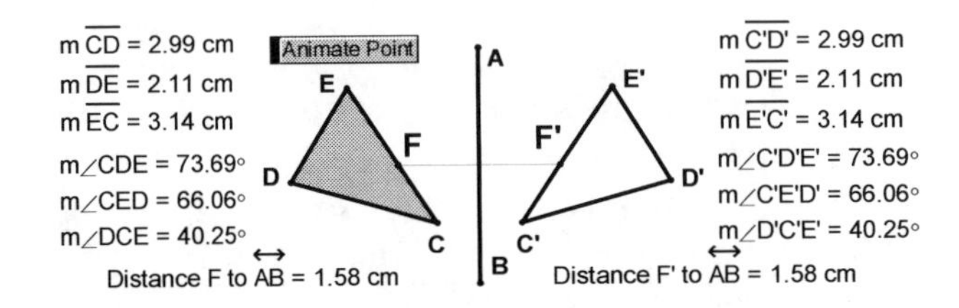

1. The distance from vertex C' to line $\overleftrightarrow{AB}$ is also 5.
2. Use the equation $4x = 28$. $x = 7$.
3. Use the equation $3x - 20 = x + 50$. Subtracting x from each side, we have $2x - 20 = 50$. Adding 20 to both sides, we have $2x = 70$ or $x = 35$. The distance from E to the line of reflection $\overleftrightarrow{AB}$ is $3x - 20 = 3(35) - 20 = 105 - 20 = 85$.

LAB #47: A TRANSLATION IN THE PLANE

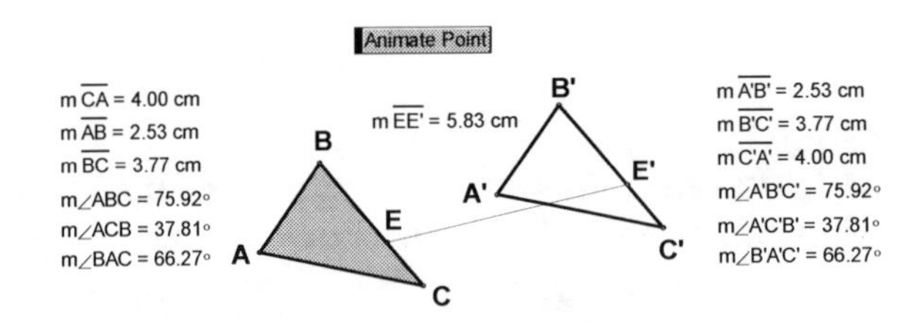

1. $m\overline{A'B'} = 1.3$ in, $m\overline{B'C'} = 2.1$ in, and $m\overline{A'C'} = 2.9$.
2. Use the equation $4x = 32$. $x = 8$.
3. Use the equation $2x + 3 = x + 5$. Subtracting x from both sides, we have $x + 3 = 5$ or $x = 2$.

LAB #48: A ROTATION ABOUT A POINT IN THE PLANE

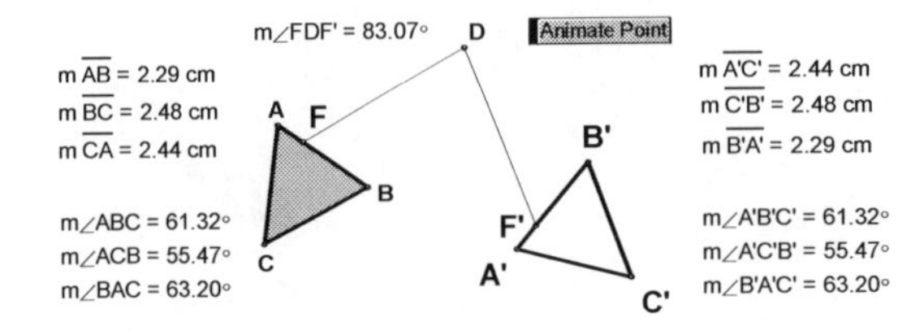

1. $m\angle A' = 63°$.
2. Use the equation $5x = 40$. $x = 8$.
3. Use the equation $3x - 3 = x + 5$. Subtracting x from each side, we have $2x - 3 = 5$. Adding 3 to each side, we have $2x = 8$ or $x = 4$.

LAB #49: DILATION ABOUT A POINT IN THE PLANE

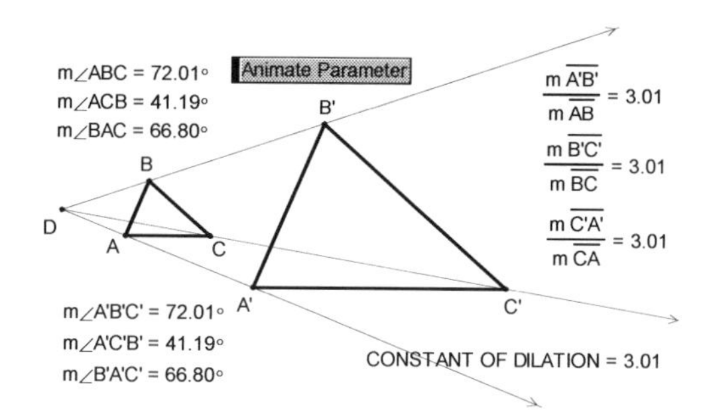

1. $m\overline{A'B'} = 20$, $m\overline{B'C'} = 28$ and $m\overline{A'C'} = 44$.
2. $m\overline{AB} = 6$.
3. Divide the image $\overline{A'B'}$ by $\overline{AB}$ to get $15 \div 3 = 5$.
4. Divide the image $\overline{A'B'}$ by $\overline{AB}$ to get $3 \div 9 = \frac{1}{3}$.

LAB #50: THE MIDSEGMENT OF A TRAPEZOID

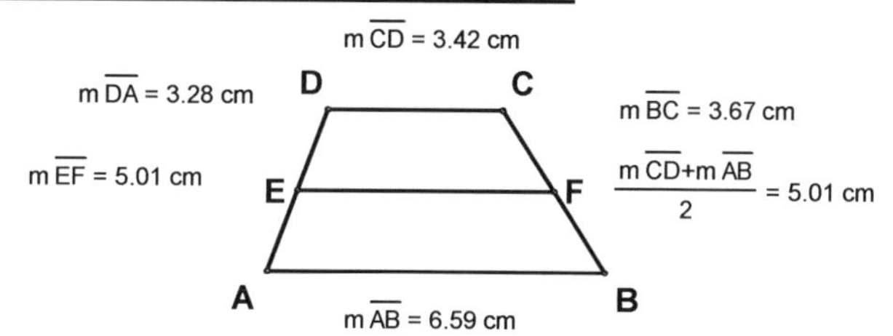

1. Taking the average of the two bases, we have EF = $\frac{12+20}{2} = 16$.
2. Let x = base AB. Use the equation $\frac{5+x}{2} = 8$. Multiplying both sides by 2, we have $5 + x = 16$, or $x = 11$.
3. Use the equation $\frac{2x+1+3x-17}{2} = 7$. Combine like terms to get $\frac{5x-16}{2} = 7$. Multiplying both sides by 2, we have $5x - 16 = 14$. Adding 10 to both sides, we have $5x = 30$ or $x = 6$. $m\overline{DE} = 2x + 1 = 2(6) + 1 = 13$. $m\overline{AC} = 4x - 11 = 4(6) - 11 = 13$.

LAB #51: THE AREA OF A SQUARE

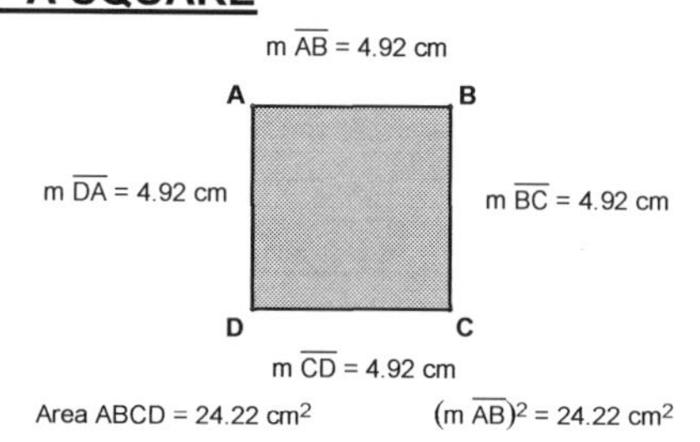

1. Use the formula $A = s^2 = 7^2 = 49$.
2. Use the formula $A = s^2 = (2x)^2 = 4x^2$.
3. Use the formula $A = s^2 = (x + 3)^2 = (x + 3)(x + 3) = x^2 + 6x + 9$.
4. Use the formula $A = s^2$ to get $36 = s^2$. Taking the square root of both sides, we have $s = 6$.
5. Use the formula $A = s^2$ to get $16x^2 = s^2$. Taking the square root of each side, we have $s = 4x$.
6. $x^2 + 10x + 25 = (x + 5)^2$. Using $s^2 = (x + 5)^2$, we have $s = x + 5$.

LAB #52: THE AREA OF A RECTANGLE

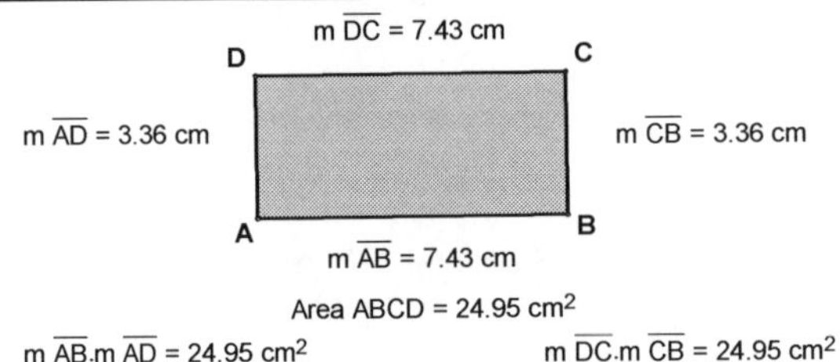

1. 78 ft^2.
2. Use the equation $6x = 87$. Dividing both sides by 6, we have $x = 14.5$.
3. Use the equation $9(2x - 3) = 63$. Dividing both sides by 9, we have $2x - 3 = 7$. Adding 3 to both sides, we have $2x = 10$ or $x = 5$. $m\overline{BC} = 2x - 3 = 2(5) - 3 = 7$.
4. Use the equation $(2x)(x) = 98$, or $2x^2 = 98$. Dividing both sides by 2, we have $x^2 = 49$. Taking the square root of both sides, we have $x = 7$. The measure of base $\overline{AB} = 2x = 2(7) = 14$. The measure of altitude $\overline{AD} = x = 7$.

LAB #53: THE AREA OF A PARALLELOGRAM

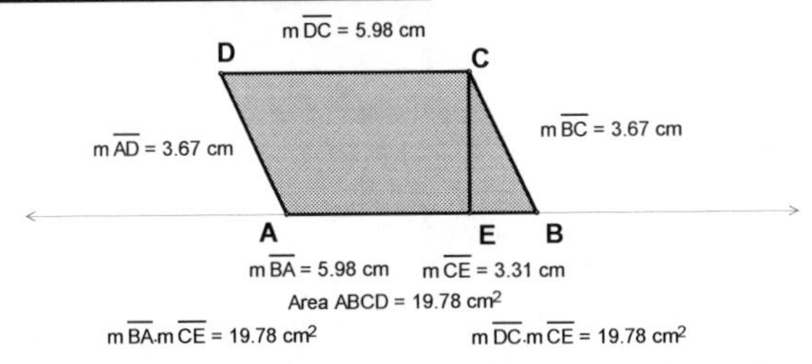

1. 105 ft^2.
2. Use the equation $12x = 60$. Then, $x = 5$.
3. Use the equation $7(x - 2) = 70$. Dividing both sides by 7, we have $x - 2 = 10$, or $x = 12$. $m\overline{AB} = x - 2 = 12 - 2 = 10$.

LAB #54: THE AREA OF A TRIANGLE

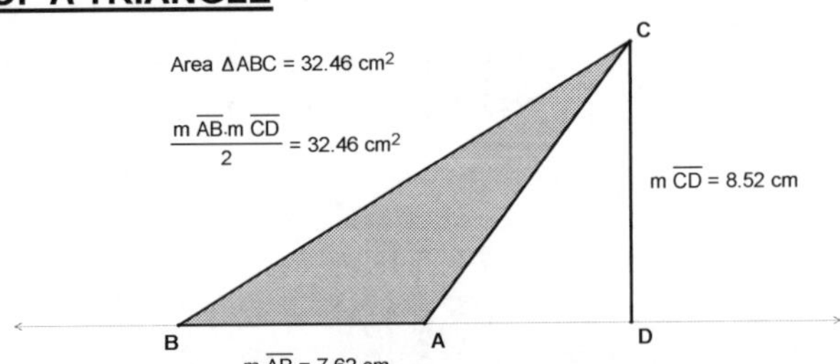

1. Use the formula for the area of a triangle $A = \frac{bh}{2} = \frac{(12)(5)}{2} = 30\ ft^2$.
2. Use the formula for the area of a triangle $A = \frac{bh}{2}$. Use the equation $40 = \frac{(10)(h)}{2}$. Simplify the fraction in this equation to get $40 = 5h$ or $h = 8$.

3. Use the formula for the area of a triangle $A = \frac{bh}{2}$. Use the equation $36 = A = \frac{(4x)(2x)}{2}$. Simplifying the fraction, we have $36 = 4x^2$. Divide both sides by 4 to get $x^2 = 9$. Take the square root of each side to get $x = 3$. $m\overline{CD} = 2x = 2(3) = 6$. $m\overline{AB} = 4x = 4(3) = 12$.

4. Use the formula for the area of a triangle $A = \frac{bh}{2}$. Use the equation $24 = \frac{(x-2)(2x)}{2}$. Reducing the fraction to lowest terms, we have $24 = x(x - 2)$. Using the distributive property, we have $x^2 - 2x = 24$. Subtracting 24 from both sides to rewrite the quadratic equation in standard form, we have $x^2 - 2x - 24 = 0$. Factoring this quadratic equation's left side, we have $(x - 6)(x + 4) = 0$. Setting each factor equal to zero, we have $x = 6$ and $x = -4$. Reject $x = -4$ since it produces a base of -6, which is impossible. So, $x = 6$ and $m\overline{CD} = 2x = 2(6) = 12$. $m\overline{AB} = x - 2 = 4$.

LAB #55: THE AREA OF A TRAPEZOID

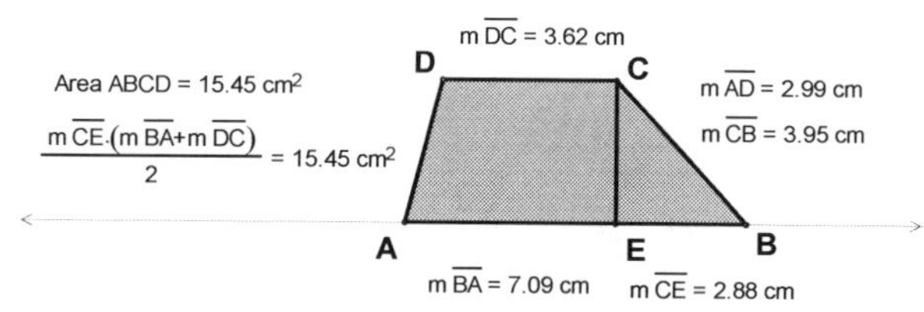

1. Use the formula $A = \frac{h(b_1 + b_2)}{2}$ to get $\frac{7(12+16)}{2} = 98$ ft^2.

2. Use the formula $A = \frac{h(b_1 + b_2)}{2}$ to get $\frac{4(6+x)}{2} = 46$. Multiply both sides by 2 to get $4(6 + x) = 92$. Using the distributive property, we have $24 + 4x = 92$. Subtracting 24 form both sides, we have $4x = 68$ or $x = 17$. $m\overline{CD} = 17$ in.

3. Use the formula $A = \frac{h(b_1 + b_2)}{2}$ to get $\frac{6(x+4+5x-6)}{2} = 48$. Multiplying both sides by 2, we have $6(x + 4 + 5x - 6) = 96$. Combining like terms within the parenthesis, we have $6(6x - 2) = 96$. Using the distributive property, we have $36x - 12 = 96$. Adding 12 to both sides, we have $36x = 108$, or $x = 3$. $m\overline{AB} = x + 4 = 3 + 4 = 7$. $m\overline{DC} = 5x - 6 = 5(3) - 6 = 9$.

LAB #56: AREA OF A SQUARE USING ITS DIAGONALS

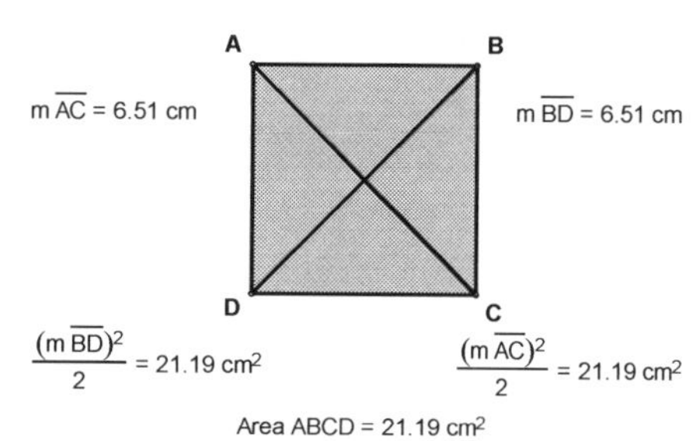

1. Use the formula $A = \frac{d^2}{2}$, we have $A = \frac{7^2}{2} = 24.5$.

2. Use the formula $A = \frac{d^2}{2}$ to get $72 = \frac{x^2}{2}$. Doubling each side, we have $x^2 = 144$ or $x = 12$. Each diagonal is 12.
3. Use the formula $A = \frac{d^2}{2}$ to get $32 = \frac{(2x)^2}{2}$. Doubling each side and squaring the 2x, we have $64 = 4x^2$. Dividing both sides by 4, we have $16 = x^2$ or $x = 4$.
Each diagonal is $2x = 2(4) = 8$.

LAB #57: DISTINGUISHING BETWEEN THE AREA AND PERIMETER OF A TRIANGLE

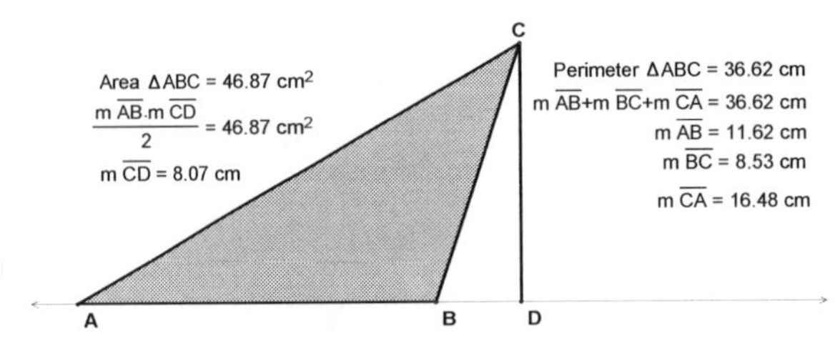

1. (a) The area is the product of the base and altitude divided by 2. This is $\frac{(10)(5\sqrt{3})}{2} = 25\sqrt{3}$ in^2. (b) The perimeter is the sum of the sides: $10 + 10 + 10 = 30$.
2. The hypotenuse can be found by using the formula $a^2 + b^2 = c^2$. $3^2 + 4^2 = c^2$, or $c = 5$. The perimeter is $3 + 4 + 5 = 12$ in. If one leg is the base, the other is the altitude since the two legs are perpendicular. The area $= \frac{(3)(4)}{2} = 6$ in^2.
3. The base AB = 16. The area is the product of the base and altitude divided by 2. This is $\frac{(16)(15)}{2} = 120$ cm^2.

LAB #58: THE RADIUS AND DIAMETER OF A CIRCLE

Radius ⊙CD = 2.91 cm   m $\overline{CD}$ = 2.91 cm

B
C
D
A

m $\overline{BA}$ = 5.82 cm   2·m $\overline{CD}$ = 5.82 cm

1. The diameter is twice the radius and is equal to 8 in.
2. The radius is one half the diameter and is equal to 6 in.
3. The radius is one half of 6x and can be written as 3x.
4. The diameter is twice the radius and can be written as 8x + 6.

LAB #59: THE CIRCUMFERENCE OF A CIRCLE AND π

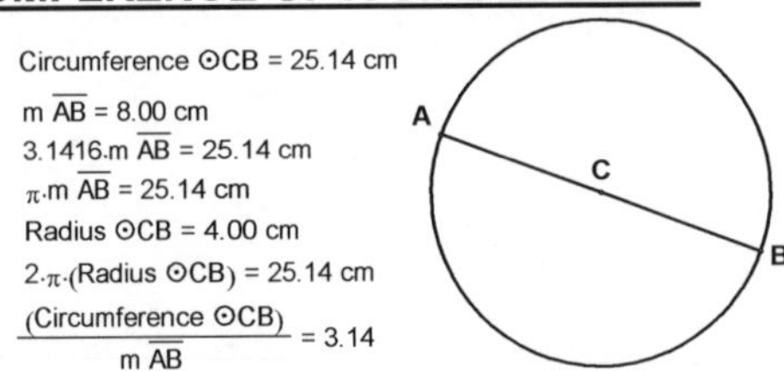

1. Use the formula $C = \pi d$. $C = 7\pi$ or $C = 22.0$.

2. Use the formula $C = \pi d$. $34 = \pi d$. Dividing both sides by π, we have $\frac{34}{\pi}$ or 10.8.
3. The amount that a wheel turns in one turn is the circumference of the wheel. Use the formula $C = \pi d$. $C = 9\pi = 28.3$.
4. Use the formula $C = \pi d$. $5x = \pi d$. Dividing both sides by π, we have $d = \frac{5x}{\pi}$.

LAB #60: THE AREA OF A CIRCLE

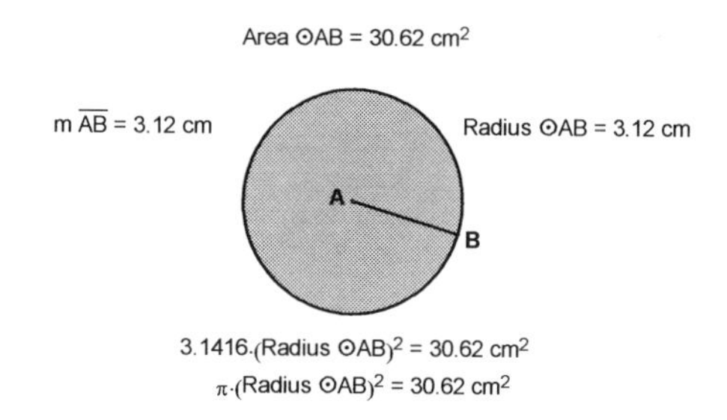

1. Use the formula $A = \pi r^2$ to get $A = \pi(3)^2$ or 9π.
2. Use the formula $A = \pi r^2$ to get $153.86 = \pi r^2$. Dividing both sides by π, we have $\frac{153.938}{\pi} = 48.999987 = r^2$. Taking the square root of each side, we have $r = 6.9999991$ or 7.
3. Use the formula $A = \pi r^2$ to get $36\pi = \pi r^2$. Dividing both sides by π, we have $r^2 = 36$ or $r = 6$.
4. Use the formula $A = \pi r^2$, $d = 24$ or $r = 12$ to get $A = \pi(12)^2 = 144\pi$.

LAB # 61: THE THREE ANGLE BISECTORS OF A TRIANGLE

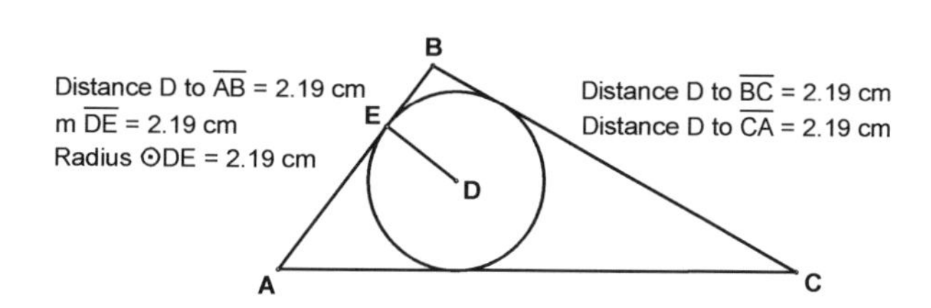

1. 7 cm.
2. Use the equation $4x = 32$. $x = 8$ in.
3. Use the equation $5x + 12 = 2x + 24$. Subtracting $2x$ from each side, we have $3x + 12 = 24$. Subtracting 12 from each side, we have $3x = 12$ or $x = 4$.
   $m\overline{DE} = 5x + 12 = 5(4) + 12 = 32$.

LAB #62: THE TANGENT RATIO IN A RIGHT TRIANGLE

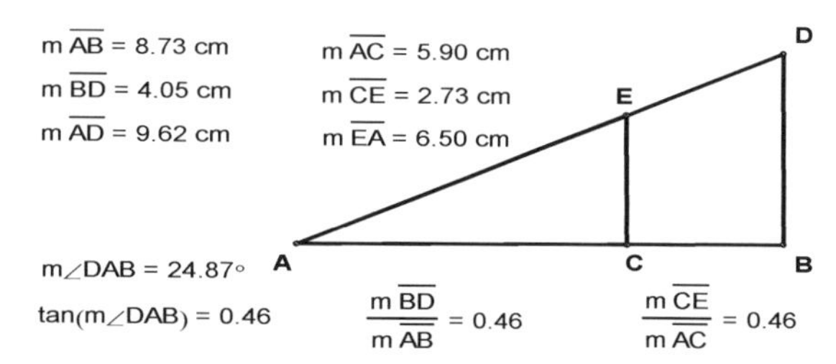

1. Using the fraction for the tangent of an angle, we have $\frac{\text{opposite leg}}{\text{adjacent leg}} = \frac{2.6}{7.2} = .36$. (Using a calculator, we also have tan 20 = .36.)

2. Use the equation tan $\angle$DAB = $\frac{\text{opposite leg}}{\text{adjacent leg}} = \frac{BD}{6}$. Using the given value for tan $\angle$DAB, we have 1.75 = $\frac{BD}{6}$. Multiplying both sides by 6, we have BD = 10.5.

3. Use the equation tan $\angle$DAB = $\frac{\text{opposite leg}}{\text{adjacent leg}}$, we have 2.68 = $\frac{4}{AB}$. Thinking of 2.68 as the fraction $\frac{2.68}{1}$, we have $\frac{2.68}{1} = \frac{4}{AB}$. Cross multiplying, we have 2.68AB = 4. Dividing both sides by 2.68, we have 1.49.

LAB #63: THE SINE RATIO IN A RIGHT TRIANGLE

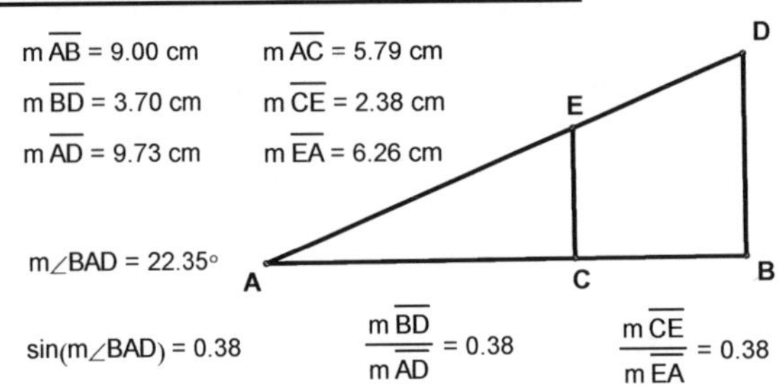

1. Using sin$\angle$DAB = $\frac{\text{opposite leg}}{\text{hypotenuse}} = \frac{5.99}{7.224}$ = .829. Using a calculator, sin$\angle$DAB = .829.

2. Using sin$\angle$DAB = $\frac{\text{opposite leg}}{\text{hypotenuse}}$, we have .7564 = $\frac{x}{7}$. Multiplying both sides by 7, we have x = 5.295.

3. Using sin$\angle$CAE = $\frac{\text{opposite leg}}{\text{hypotenuse}}$, we have .8547 = $\frac{5}{x}$. Multiplying both sides by x, we have .8547x = 5. Dividing both sides by .8547, we have 5.85.

LAB #64: THE COSINE RATIO IN A RIGHT TRIANGLE

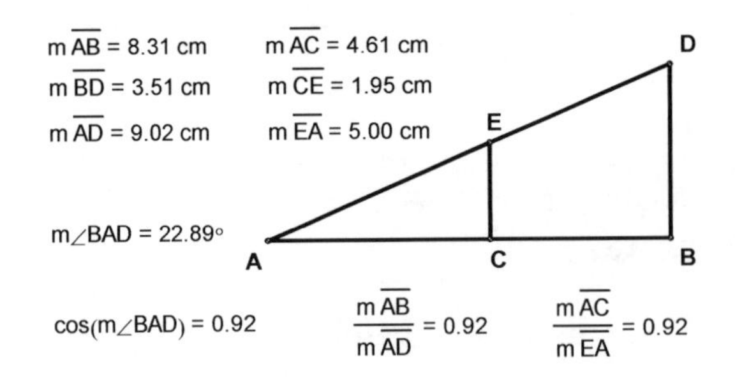

1. Using cos$\angle$DAB = $\frac{\text{adjacent leg}}{\text{hypotenuse}} = \frac{3.75}{6.38}$ = .588. Using a calculator, cos$\angle$DAB = .588.

2. Using cos$\angle$DAB = $\frac{\text{adjacent leg}}{\text{hypotenuse}}$, we have .3456 = $\frac{x}{9}$. Multiplying both sides by 9, we have x = 3.11.

3. Using cos$\angle$CAE = $\frac{\text{adjacent leg}}{\text{hypotenuse}}$, we have .6801 = $\frac{4.76}{x}$. Multiplying both sides by x, we have .6801x = 4.76. Dividing both sides by .6801, we have 7.

LAB #65: NAMING A POINT BY USING ITS COORDINATES

A: (3.00, 4.00)

$x_A = 3.00 \quad y_A = 4.00$

1. The x-coordinate of point B is 5. The y-coordinate is -7.
2.

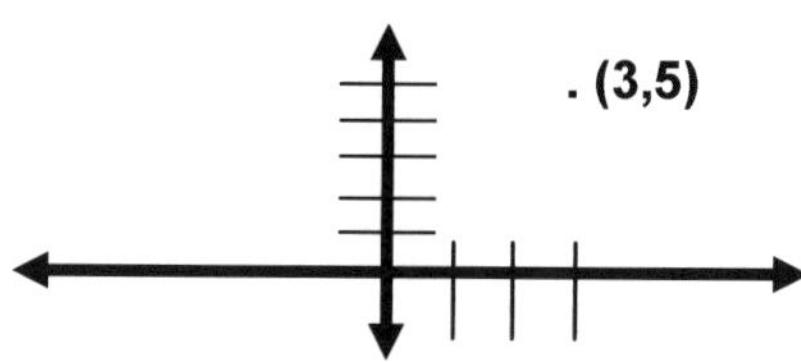

3.

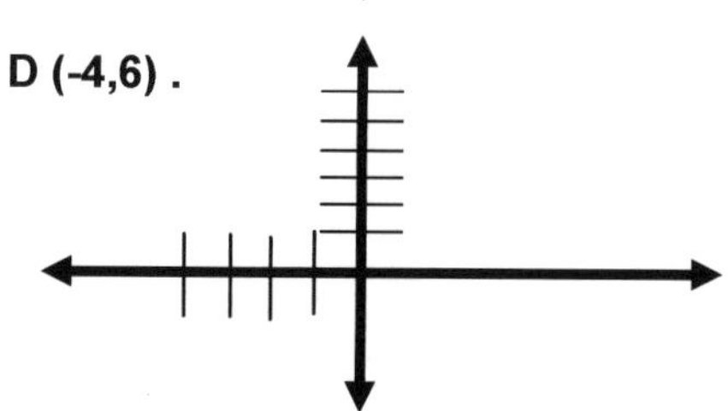

4.

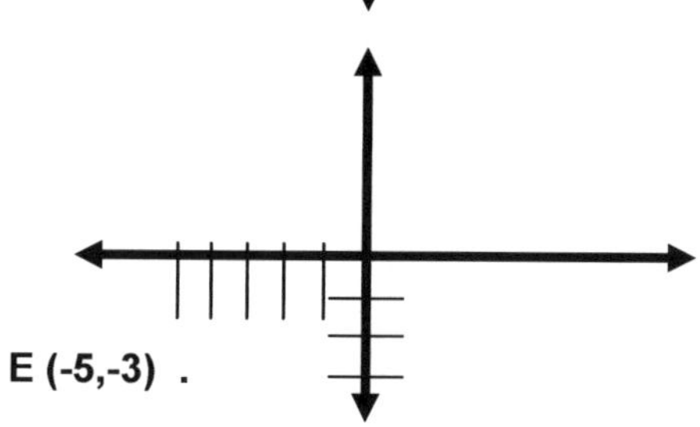

LAB #66: THE SLOPE OF A LINE

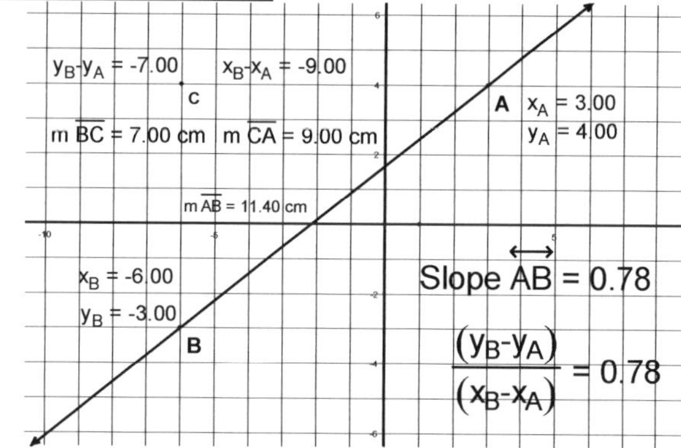

1. Use the formula $\dfrac{y_2 - y_1}{x_2 - x_1} = \dfrac{6-4}{4-3} = 2.$

2. Use the formula $\dfrac{y_2 - y_1}{x_2 - x_1} = \dfrac{5-3}{4-(-2)} = \dfrac{2}{6} = \dfrac{1}{3}.$

3. Use the formula $\frac{y_2 - y_1}{x_2 - x_1} = \frac{(-7)-4}{8-(-3)} = \frac{-11}{11}$ **= -1.**

LAB #67: THE SLOPES OF PARALLEL LINES

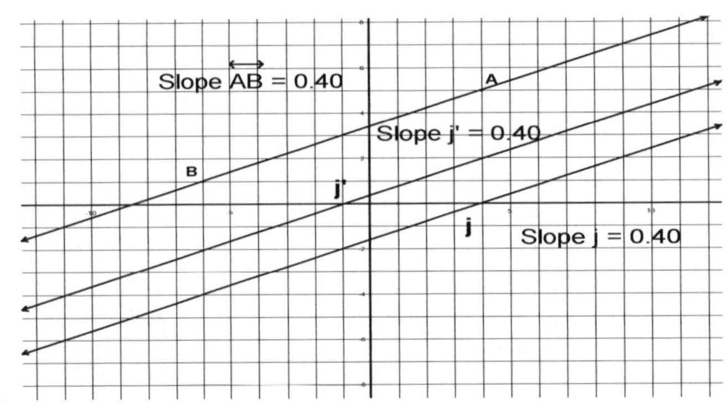

1. **The slope is 2.**
2. **Since the lines are parallel, the slopes are equal. Use the equation 2x = 5 or x = 2.5.**
3. **Since the lines are parallel, the slopes are equal. Use the equation 2x + 2 = x + 4. Subtracting x from each side, we have x + 2 = 4 or x = 2. Each slope is 2(2) + 2 = 6.**

LAB # 68: THE SLOPES OF PERPENDICULAR LINES

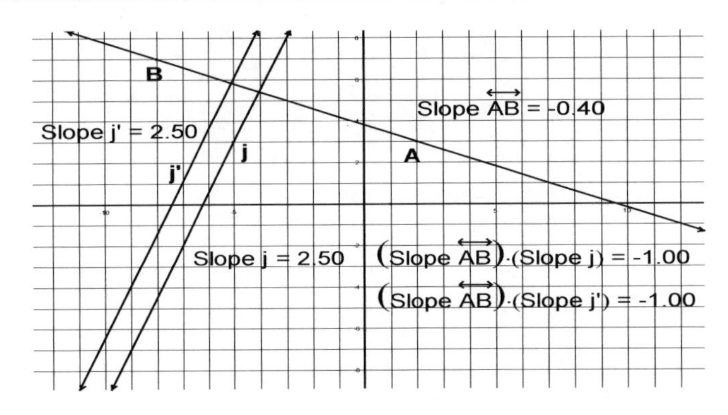

1. **Multiplying the two slopes, we have (2)(-0.5) = -1.**
2. **The slope is the negative reciprocal of 4,** $-\frac{1}{4}$**.**
3. **The slope is the negative reciprocal of x,** $-\frac{1}{x}$**.**

LAB # 69: THE SLOPE AND Y-INTERCEPT OF A LINE

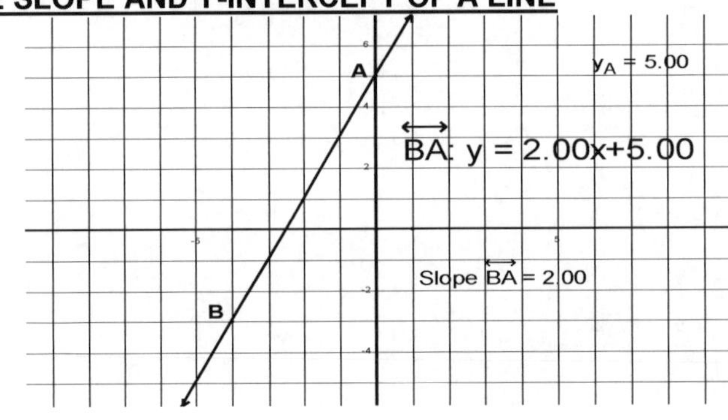

1. The slope of $y = 2x + 4$ is 2 and the y-intercept is 4.
2. The slope of $y = 3x - 7$ is 3 and the y-intercept is -7.
3. Using the standard form $y = mx + b$, we have $y = 4x + 6$.
4. Using the standard form $y = mx + b$, we have $y = x - 2$.

LAB #70: THE EQUATION OF A VERTICAL LINE ON THE COORDINATE PLANE

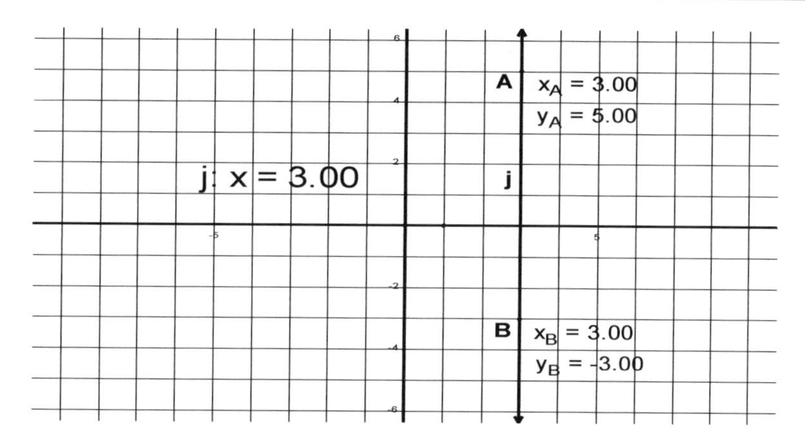

1. The value of every x-coordinate on this line is -2.

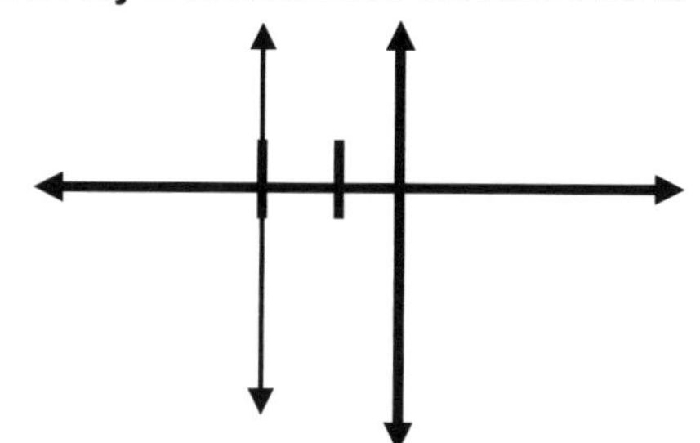

2. The value of every x-coordinate on this line is 4.

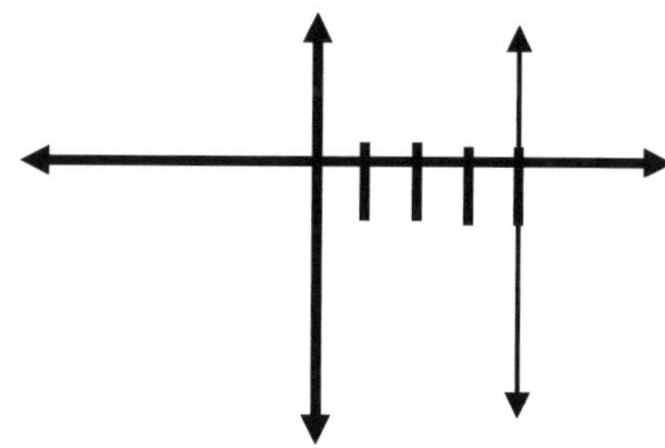

3. The value of every x-coordinate on this line is 0. This is the y-axis.

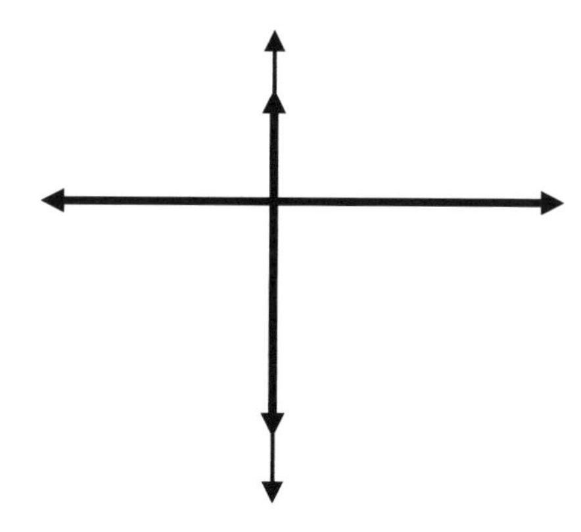

LAB #71: THE EQUATION OF A HORIZONTAL LINE ON THE COORDINATE PLANE

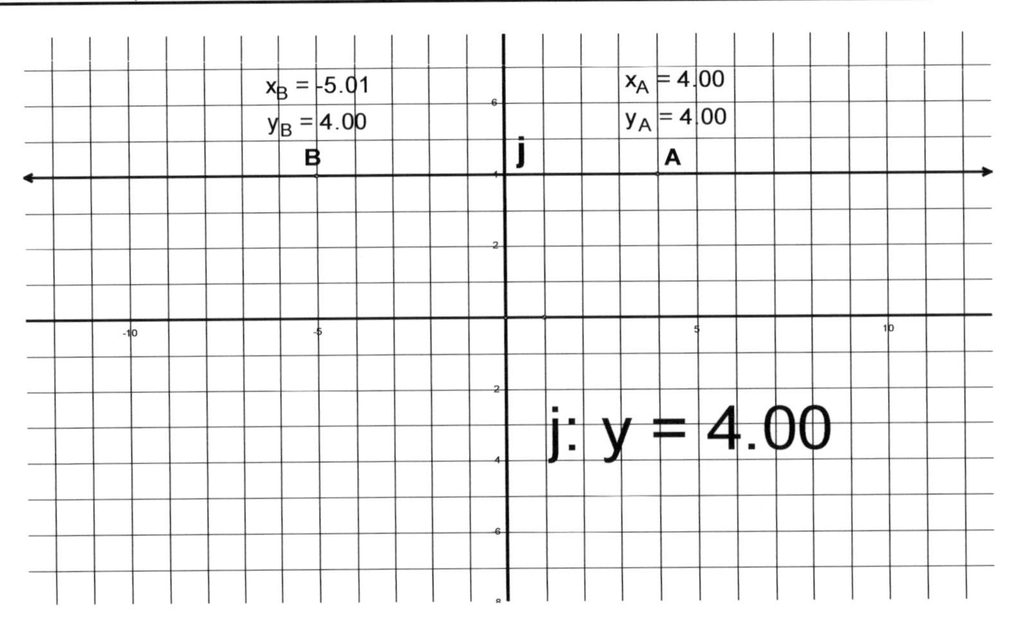

1. The value of every y-coordinate on this line is 4.

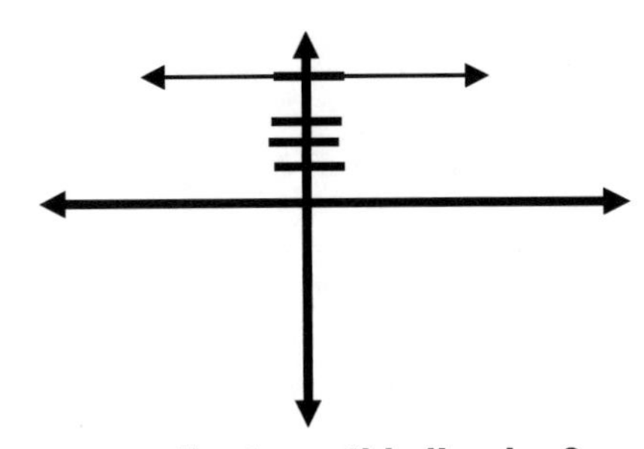

2. The value of every y-coordinate on this line is -2.

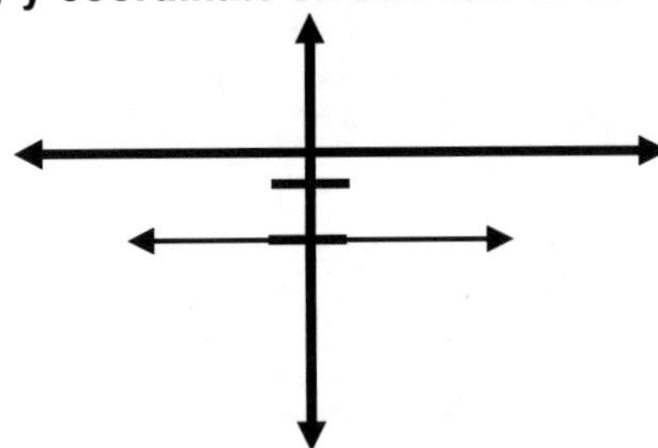

3. The value of every y-coordinate on the x-axis is 0.

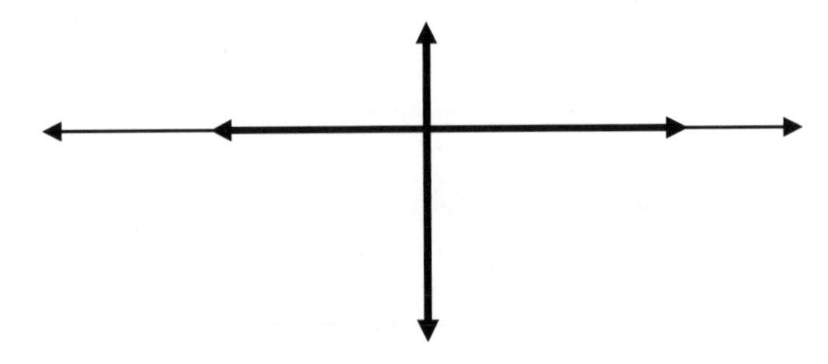

LAB #72: THE COORDINATES OF THE MIDPOINT OF A LINE SEGMENT

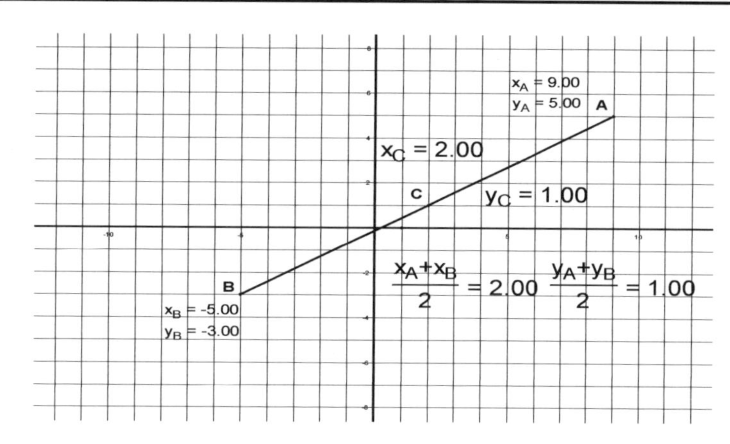

1. Use the midpoint formula to get $(\frac{1+3}{2}, \frac{5+9}{2})$ = (2, 7).
2. Use the midpoint formula to get $(\frac{3+7}{2}, \frac{4+8}{2})$ = (5, 6).
3. Use the midpoint formula to get $(\frac{4+x}{2}, \frac{1+y}{2})$ = (6, 5). Equating the two x-coordinates, we have $\frac{4+x}{2}$ = 6. Multiplying both sides of the equation, we have 4 + x = 12 or x = 8. Equating the two y-coordinates, we have $\frac{1+y}{2}$ = 5. Multiplying both sides of the equation, we have 1 + y = 10 or y = 9. So the coordinates of B are (8, 9).

LAB #73: THE DISTANCE BETWEEN THE ENDPOINTS OF A HORIZONTAL LINE SEGMENT

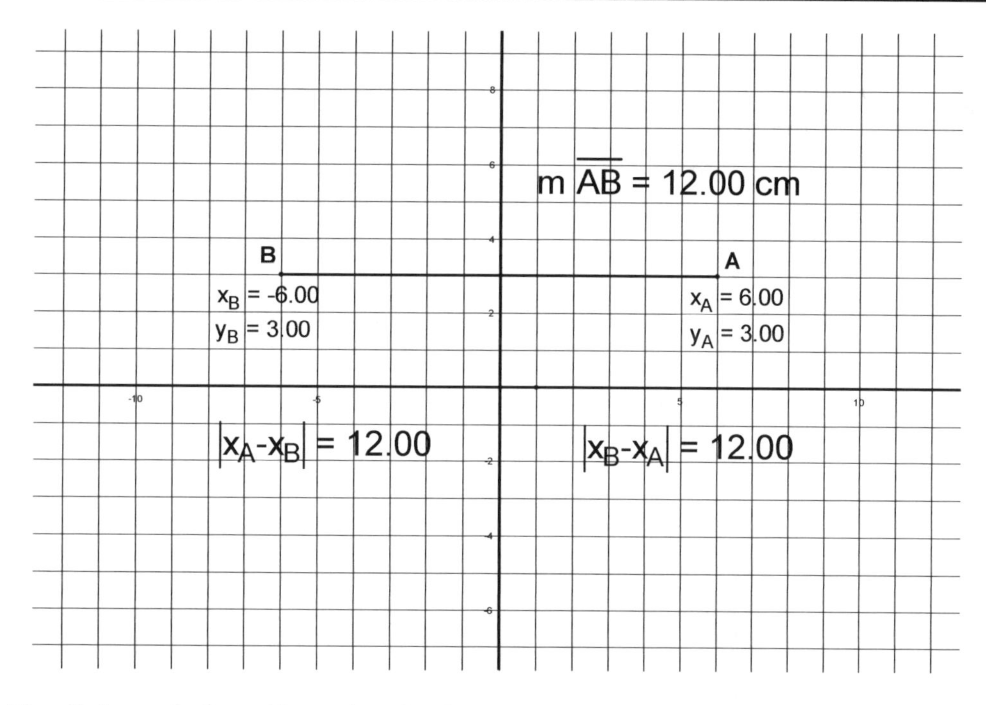

1. The distance is found by using the formula | 7 – 2 | = 5.
2. The distance is found by using the formula | 5 – (-2) | = 7.
3. Using the distance formula, we have | x – 4 | = 6. This could either be 10 or -2.

LAB #74: THE DISTANCE BETWEEN THE ENDPOINTS OF A VERTICAL LINE SEGMENT

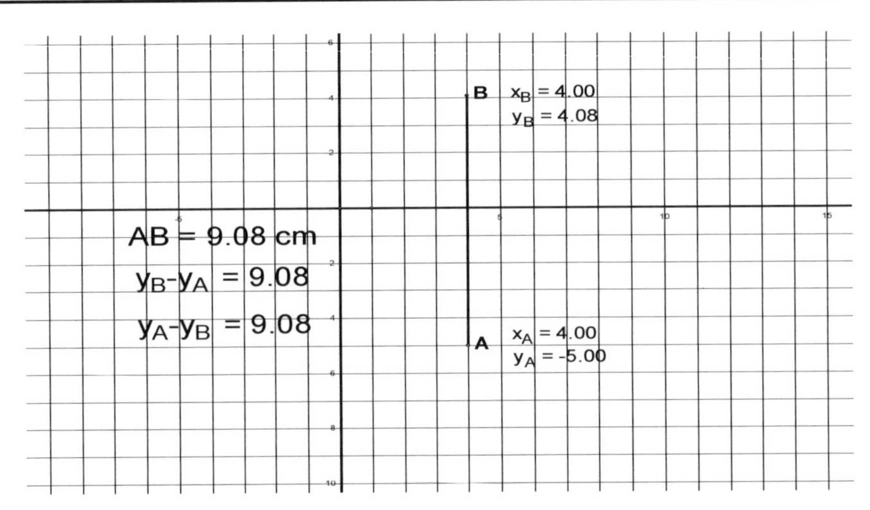

1. **Use the formula AB = $|y_2 - y_1|$ to get $|7 - 2| = 5$.**
2. **Use the formula BD = $|y_2 - y_1|$ to get $|4 - (-3)| = 7$.**
3. **Use the formula AB = $|y_2 - y_1|$ to get $|y - 2| = 9$. y could be either equal to 11 or -7.**

LAB #75: THE DISTANCE BETWEEN ENDPOINTS OF ANY LINE SEGMENT

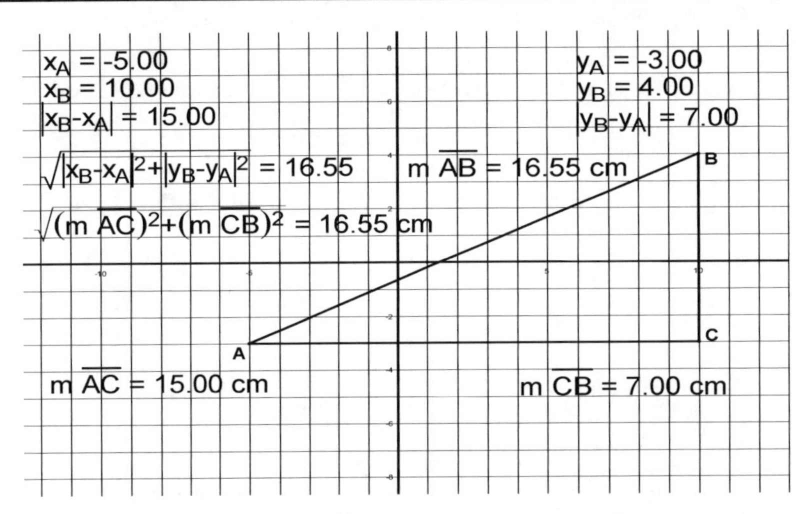

1. **Use the formula to find the distance between two points**
   $D = \sqrt{(x_2 - x_1)^2 + (y_2 - y_1)^2}$. We have AB = $\sqrt{(9-5)^2 + (1-(-2))^2} = \sqrt{(4)^2 + (3)^2} = 5$.
2. **Use the formula to find the radius from the center of the circle to a point on the circle**
   $D = \sqrt{(x_2 - x_1)^2 + (y_2 - y_1)^2}$. We have radius = $\sqrt{(7-1)^2 + ((-8)-3)^2} = \sqrt{(6)^2 + (-11)^2} =$
   $\sqrt{36+121} = \sqrt{157} = 12.5$.
3. **First use the formula to find side AB = $\sqrt{(x_2 - x_1)^2 + (y_2 - y_1)^2}$.**
   We have AB = $\sqrt{(9-1)^2 + (4-1)^2} = \sqrt{(8)^2 + (3)^2} = \sqrt{64+9} = \sqrt{73}$. Then use the formula to find side BC = $\sqrt{(9-1)^2 + (4-7)^2} = \sqrt{8^2 + (-3)^2} = \sqrt{64+9} = \sqrt{73}$. Since two sides of the triangle have the same measure, the triangle is isosceles.

23861815R00117

Made in the USA
San Bernardino, CA
03 September 2015